AF764281

# BRANDON

## Geographical Perspectives on the Wheat City

**Edited by John Welsted**
**John Everitt**
**and Christoph Stadel**

CANADIAN PLAINS RESEARCH CENTER
UNIVERSITY OF REGINA
1988

© John Welsted, John Everitt, and Christoph Stadel

**Canadian Cataloguing in Publication Data**

Main entry under title:

Brandon: geographical perspectives on the wheat city

1st ed.

Includes bibliographical references and index.
ISBN 0-88977-052-2

1. Brandon (Man.) — Description. I. Welsted, John E.
II. Everitt, John C. III. Stadel, Christoph.
IV. University of Regina. Canadian Plains Research Center.

FC3399.B73B74 1988 917.127'3 C88-098024-9
F1064.5.B73B74 1988

Printed in Canada by D.W. Friesen & Sons Ltd.

with the compliments of the editors

John Welsted

John Everitt

# BRANDON

# CONTENTS


| | | |
|---|---|---|
| | **Figures** | vi |
| | **Plates** | viii |
| | **Tables** | ix |
| | **Foreword**<br>John Warkentin | xi |
| | **Preface**<br>John Welsted, John Everitt and Christoph Stadel | xv |
| One | **Brandon in its Setting**<br>John Welsted, John Everitt and Christoph Stadel | 1 |
| Two | **Brandon and the Assiniboine River**<br>John Welsted | 9 |
| Three | **The Climate of Brandon and its Surroundings**<br>Rod McGinn | 37 |
| Four | **Spatial Growth of Brandon**<br>John Everitt and Christoph Stadel | 61 |
| Five | **Population Structure and Social Areas**<br>Alison Gill | 89 |
| Six | **Brandon's Manufacturing Sector**<br>Paul Curtis | 105 |
| Seven | **Downtown Brandon**<br>Christoph Stadel and John Everitt | 123 |
| Eight | **The Urban Fringe**<br>John Everitt and Christoph Stadel | 151 |
| Nine | **Urban Planning in the City of Brandon**<br>Ron Tompkins | 177 |
| Ten | **Brandon's Role as a Service Centre in Southwestern Manitoba**<br>Christoph Stadel and John Everitt | 195 |
| Eleven | **Conclusion and Outlook**<br>John Welsted, John Everitt and Christoph Stadel | 221 |
| | **Notes on Contributors** | 224 |

## FIGURES


1:1 Location of the city of Brandon 2
1:2 Physical features of southwestern Manitoba (modified from Bird, 1972) 3
2:1 Cross profiles of the Assiniboine Valley at Brandon 10
2:2 The Assiniboine Basin 12
2:3 Location of natural and human-made features in the Assiniboine Valley at Brandon 13
2:4 Cross channel profiles of the Assiniboine River at Brandon 16
2:5 Changes in the course of the Assiniboine River between 1881 and 1987 30
2:6 Suggested diversions of the Assiniboine River at Brandon 33
3:1 The seasons for Brandon 39
3:2 Climatic diagram for Brandon 39
3:3 Radiation and sunshine hours for Brandon 43
3:4 The climatological water balance for Brandon 56
4:1 Brandon's population growth: 1881-1978 70
4:2 Spatial growth of the built-up area of Brandon: 1881-1981 71
4:3 A model of Brandon's social areas: 1883-1914 74
4:4 Residential subdivisions of Brandon: 1954-1982 82
4:5 A conceptual model of suburban development in Brandon 85
5:1 Changes in Brandon's age structure: 1971-1981 90
5:2 A comparison of the age structures for Brandon and Winnipeg: 1971 90
5:3 Spatial variance in "family status" in Brandon 93
5:4 Spatial variance in "household type" in Brandon 96
5:5 Spatial variance in "ethnic status" in Brandon 97
5:6 Spatial variance in "economic/educational status" in Brandon 100
5:7 Spatial variance in "age/marital status" in Brandon 101

| | | |
|---|---|---|
| 6:1 | Number of manufacturing plants and number of employees engaged in manufacturing in Brandon: 1923-1980 | 108 |
| 7:1 | Schematic representation of the Central Business District (CBD) of Brandon | 127 |
| 7:2 | Land use in Downtown Brandon in 1986 | 131 |
| 8:1 | The Brandon Area Planning District | 153 |
| 8:2 | Boundary changes for the city of Brandon | 158 |
| 8:3 | Land use on Brandon's northern fringe: 1979 | 161 |
| 10:1 | Schematic representation of Brandon's location within the network of service centres in southwestern Manitoba | 197 |
| 10:2 | Population change in the service centres of southwestern Manitoba: 1971-1981 | 204 |
| 10:3 | Brandon's service areas | 206 |

# PLATES


| | | |
|---|---|---|
| 2:1 | Logs stored on the Assiniboine River during Hanbury's "big drive" of 1910 | 18 |
| 2:2 | Ferry crossing the Assiniboine at about Third Street in 1882 | 18 |
| 2:3 | The 1922 spring flood | 26 |
| 4:1 | Rosser Avenue in 1883 | 64 |
| 4:2 | Pacific Avenue in 1885 | 67 |
| 4:3 | Brandon's Central Business District in 1919 | 68 |
| 4:4 | The 700 Block of Rosser Avenue in 1919 | 72 |
| 5:1 | The Assiniboine Gardens area of Brandon | 98 |
| 6:1 | The junction of Eighteenth Street and Richmond Avenue in 1960 | 112 |
| 6:2 | Southeastern Brandon in 1987 | 117 |
| 7:1 | Aerial view of Downtown Brandon from the northwest | 128 |
| 7:2 | Rosser Avenue in 1987 | 133 |
| 7:3 | Tenth Street in 1987 | 134 |
| 7:4 | Aerial view of Brandon from the south in 1987 | 135 |
| 7:5 | Western Brandon in 1987 | 136 |
| 8:1 | Brandon's northern fringe | 164 |

# TABLES

4:1 Value of Building Permits in Brandon: 1904-1944 69
5:1 Summary of Varimax-Rotated Factor Loadings 92
6:1 Percentage of Brandon's Labour Force Engaged in Manufacturing Compared to That in Canada 109
6:2 Percentage Changes in Brandon's Labour Force Engaged in Manufacturing Compared to Those in Canada 109
6:3 Percentage Changes in Brandon's Total Labour Force Compared to Those in Canada 109
6:4 Brandon's Basic Industries 111
7:1 Nonresidential Floor Space Within the Central Business District of Brandon 126
7:2 Land Uses Within the Central Business District of Brandon 137
10:1 Distribution of Population in Southern Manitoba, by Settlement Classes: 1971 201
10:2 Population and Population Change of Centres in Southwestern Manitoba: 1971-1981 202

# FOREWORD

*John Warkentin*

I grew up in the 1930s in the Red River Valley, the flattest region in all of North America. During the Depression travel was not that easy. I had a vague sense that there was a more expansive land to the west, and a romantic vision of a more freewheeling, exciting society beyond the Manitoba Escarpment.

Years later I taught high school physics and chemistry in Gilbert Plains, just north of the Riding Mountains. Now I heard more specifically about Brandon. A representative from Brandon College, Walter Dinsdale, came round to invite our pupils to attend that institution. Now I began to think of Brandon as a significant cultural centre, somewhere across the mountain beyond Clear Lake.

A few years afterward, in the late 1950s, when I was teaching geography at the University of Manitoba, I often did field work in the Brandon area. Finally, I came to know the Brandon region at first hand.

What then is this region like, as seen by a geographer? Many factors go to make up a region. I do not intend to list them here, except to say that to identify and know a region, just as a nation, it is absolutely vital to understand the experiences of the people who live there, and who have lived there in the past.

The west as a whole, from Winnipeg to the Rockies, is an area where a transformation took place in the nineteenth century. We will only look at the Brandon region. The great change which occurred was a shift from a life of motion, where constant movements of buffalo, Indians, Métis and traders sweeping over the plains prevailed, to that of the life of the individual settler, rooted on his own homestead. Manufactured goods were shipped to the settler, and his wheat moved in turn to world markets, but he himself was fixed in a sharply delimited place.

During the period of the region in motion, we can already see something distinctive about the Brandon area. The wooded Riding and Turtle Mountains and the deep valleys of the Assiniboine, Souris, Pembina, and Little Saskatchewan Rivers were widely recognized landmarks on the bare plains. When fur traders assigned to prairie posts went to visit the Mandan Indians on the Missouri River they travelled south from the Assiniboine, and they used prominent features such as the Turtle Mountains to guide them. The famous explorer David Thompson took such a winter break in 1797, and one of the finest descriptions of a blinding blizzard that I know is the account of his journey through this region to the Mandans. We get there perhaps too vivid a sense of this land in winter. During the fur trade era the Assiniboine River was a fine base for posts, and Riding Mountain to the north a hunting area. On the slopes of Turtle Mountain to the south a few Métis built log cabins in the 1870s and tilled some plots.

But then a revolution came to this rolling land, with its entrenched valleys, forested slopes, bluffs and natural meadows. Surveyors, settlers, railroads, towns and new institutions rushed in. A new society and geometry were imposed on the land. Indians and Métis were pushed into a few corners. Even the Riel Rebellion farther west on the Saskatchewan River was short lived.

In this region, the main story now became that of community development. It is this experience which has shaped this area, which remains with the inhabitants, and makes it a recognized region within Manitoba today.

First accounts often painted an awe-struck picture of the plains — the kind of thing I had imagined when I was growing up. These vast, still unfenced spaces and splendid skies were thought to comprise one of the most sublime landscapes on this continent, and had a deep emotional impact. They touched people. We catch something of this sense of exhilaration in getting to know this spacious new country from the animal tales of Ernest Thompson Seton set in the Sand Hills of the Carberry area. But he still emphasized space and mobility, not farming.

As it happens, this is one of the most agriculturally favoured regions in the Prairies, and the need for technological adaptations to help establish farms in a new environment was not as great here as elsewhere in the west. The hard truth, however, is that it still was a formidable challenge to convert the buffalo grazing lands into farms. It was a grinding experience — failures were recorded as well as ultimate successes. The thrill of the open plains was replaced by the attachment to an individual quarter or half section, and the deep inner satisfaction of creating a farmstead. And the new life brought a new landscape. René Dubos talks of the wooing of earth, and that is what took place. It was not just a matter of growing wheat. Farmsteads were planned; shelter belts, shrubs, and gardens planted. An attractive, domestic landscape of closely spaced farmsteads to which man could relate was created. The settlers had a profound and legitimate pride in this act of creation. That landscape is one of the remarkable accomplishments of this region, too little recognized.

Marching in lock step with the homesteaders were townspeople, founding dozens of small trading centres even as the farmers occupied the land. This is where the city of Brandon got its identity, because it became the organizing centre of the region. City and region cannot be separated, and Brandon was the undisputed urban centre of western Manitoba. It was much more than a trading centre. Experiences were shared, ideas exchanged, and a local regional consciousness was forged out of the common endeavour of transforming a land.

With the coming of the automobile and of highways Brandon's role as a crossroads and regional centre became clearer. The Canadian Pacific Railway (CPR) corridor had given the Brandon region an effective east to west dimension, but Highway 10 and other roads firmly drew in towards Brandon all the country up to Riding Mountain in the north, and to Turtle Mountain in the south, just as those two mountains had bracketed the region in the fur trade era.

Within Canada, the remarkable thing about a city the size of Brandon is that basically it has remained an agricultural centre. Neither great mineral resources nor manufacturing have propelled it forward. Whatever its ambitions, alternatives are limited, and it has had to remain true to its agricultural origins.

However, any assumption of placidity or lack of change is most deceiving. Brandon and its region have been swept by waves of change. In the 1920s and 1930s there occurred the great conversion from horse power to the internal combustion engine, the Great Depression and the impact of drought. Many farmers were forced from the land, and eventually even community centres such as rural churches disappeared. Nevertheless, though greatly thinned out, the basic farm and town structure has remained in place. Nor does life remain stable or easy. From what has happened to grain prices this year (1987) it is as clear as ever that no grain-growing region in the world is isolated.

Other fundamental changes have occurred. In the decades after World War II there has been a shift in the composition of the population. Internal migration is characteristic of Canada. In this region many farms originally settled by people of British origin have been taken over by farmers from other cultural groups, often coming from adjacent areas in Manitoba. Thus multiculturalism has increasingly come to the region. Nowhere is this more noticeable than in the reemergence of Native people, in their new and stronger role within the larger society.

Despite such continuing changes, the sense of place remains secure in western Manitoba. People have regional roots, feelings of community which come out of common problems and common achievements.

This is where the city of Brandon and Brandon University are so vital — as cultural centres in the widest sense, and as regional voices. Regional artists and the creative life in a university can help people to understand themselves and their region. This is no mean task because there is a strong geographical similarity or gradient across the Prairies, which the Brandon region has to arrest, and very potent metropolitan forces which tend to dominate and homogenize our lives. This is where the region provides a feeling of attachment. It is at a scale which we can comprehend, and in which we feel at home.

# PREFACE

*John Welsted, John Everitt and Christoph Stadel*

The idea for this book came from a desire to commemorate the twenty-fifth anniversary of the establishment, in 1962, of the Department of Geography at Brandon College (which became Brandon University in 1967). Over the years, the instructors in the department have been collecting information about the geography of the city of Brandon and its surroundings and it seemed an appropriate time to put some of it into print in a consolidated form. Consequently, several people were invited to submit articles to fit into the theme, "Brandon: Geographical Perspectives." Each of the essays is an original contribution written specifically for this volume. The book is aimed at the general reader who is interested in geography and consequently we have limited the use of technical terms and abbreviations wherever possible.

Apart from acknowledging the contributions of the authors we would like to recognize the help of a few other individuals and groups. Ms. Christine Debattista, secretary to the Department of Geography, typed the whole manuscript several times. The maps and diagrams were produced by the Brandon University Cartographics Laboratory staff under the supervision of Mr. James Darlington. We received funding from the Manitoba Heritage Federation and from a Social Sciences and Humanities Research Council grant to Brandon University. Mr. Brian Mlazgar of the Canadian Plains Research Center, University of Regina, edited the manuscript.

Finally, in that this book is written to celebrate the twenty-fifth anniversary of the founding of the Department of Geography at Brandon University, it is fitting that it should be dedicated to Dr. John Tyman who was the first geography instructor and founding chairman of the department which owes its growth and success to his dedication and perseverance in the early years.

# ONE
# Brandon in its Setting

*John Welsted*

The city of Brandon, the second city in Manitoba in terms of size and importance, is located in the eastern Prairies on the Assiniboine River two hundred kilometres west of Winnipeg (Figure 1:1). It is the economic, cultural and educational centre of southwestern Manitoba — a geographical region which is roughly bounded by the Canada-United States border in the south, the Manitoba-Saskatchewan border in the west, and the Manitoba Escarpment in the east, northeast and north. The escarpment, which extends from the North Dakota border to the Wapawekka Hills in north-central Saskatchewan, is a convenient physiographic boundary for south-western Manitoba although it is broken into separate upland units by a series of embayments, the largest of which is occupied by the Assiniboine River which meanders across southwest and south-central Manitoba to join the Red River at Winnipeg.

Brandon is no exception to the rule that a city cannot be considered in isolation from the region in which it is located. A combination of topography, soils, natural vegetation and climate resulted in southwestern Manitoba becoming one of the great agricultural regions of Canada. Early in the city's history, wheat grown on the soils of the surrounding area was transported to Brandon for shipment so that it soon became known as "The Wheat City." Today, one of the city's main functions remains that of a service centre for the surrounding agricultural region. Its fortunes, in the past and at present, are closely related to events, both natural and human-induced, within the Prairies and more specifically within the geographical region of southwestern Manitoba.

Although most of the topography of southwestern Manitoba is a relic of the Pleistocene Ice Age, the Manitoba Escarpment predates this event. During preglacial times the Cretaceous sediments which underlie much of the Prairies were down-folded into a gentle syncline, the eastern edge of which is now the escarpment. "The escarpment is an erosional feature, a rim of truncated Cretaceous

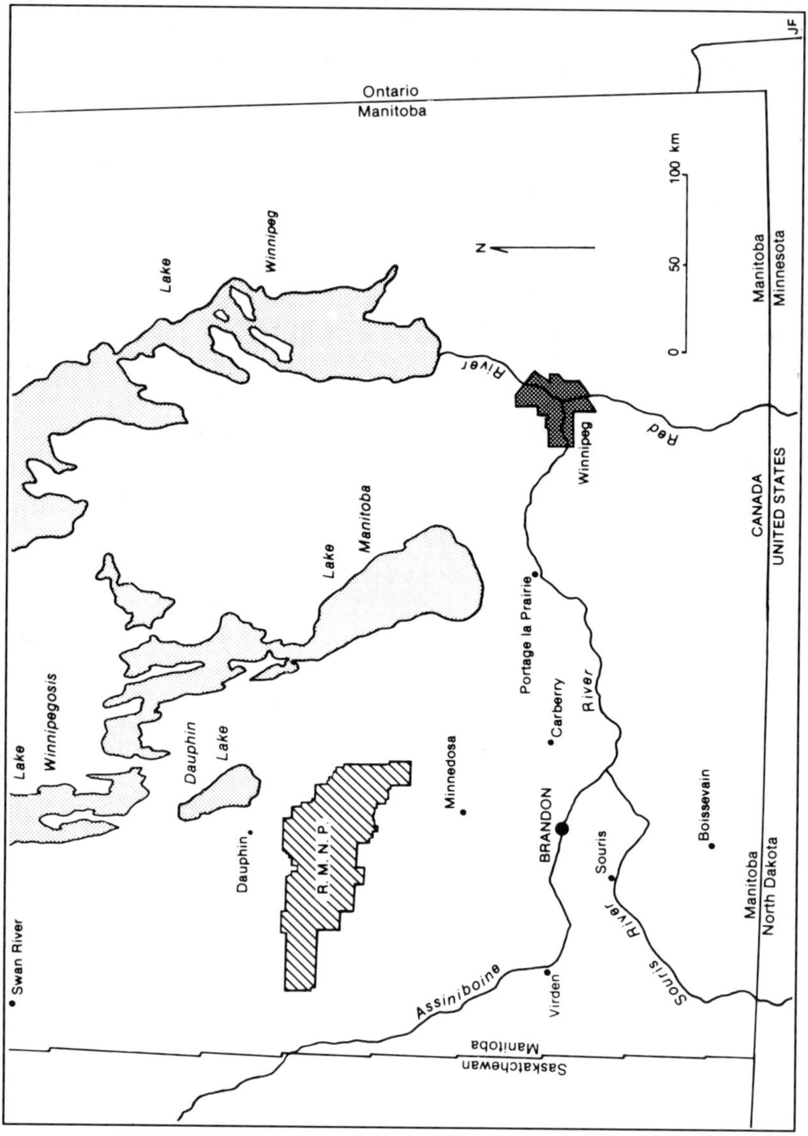


FIGURE 1:1
*Location of the city of Brandon*

rocks that once extended further to the east" (McNeil 1987:2). Eastward flowing streams dissected the escarpment to produce the "mountains" — Pembina Mountain, Riding Mountain, Duck Mountain and the Porcupine Hills which form the northeastern boundary of the physical region of southwestern Manitoba (Figure 1:2).

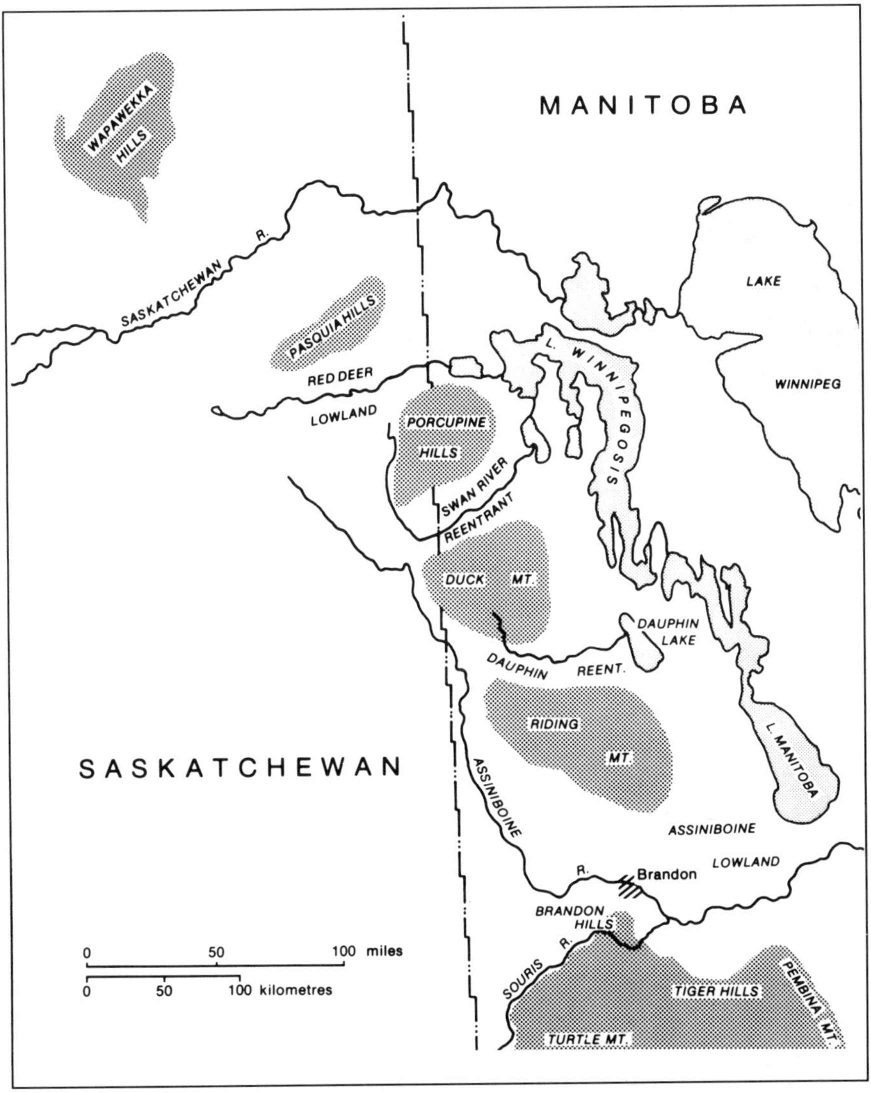


FIGURE 1:2
*Physical features of southwestern Manitoba (modified from Bird, 1972)*

To say that the Ice Age had a profound effect on the topography of southwestern Manitoba is an understatement. During the last stage — the Wisconsinan Glacial Period — ice covered all of Manitoba, eventually disappearing about twelve thousand years ago and leaving behind a series of glacial and glacially related deposits. Much of the region is covered by dead ice moraine deposited by the decaying ice sheet as it melted (Teller 1984). The resulting topography is undulating or gently rolling, with many undrained, irregularly distributed, shallow depressions (sloughs). End moraine deposited at the stationary edge of the ice sheet (Elson 1959) produced a greater relief (up to seventy-five metres) and is less suitable for farming than dead ice moraine; the closest example to Brandon is the Brandon Hills, sixteen kilometres south of the city. Glaciofluvial sediments were deposited by water in gaps and holes under and within the ice. One of the resulting landforms is an esker represented by a ridge at the eastern end of the Brandon Hills, and by the Arrow Hills northwest of the city (Welsted and Young 1980).

The Assiniboine Valley, in which Brandon is located, is an indirect product of the Ice Age, although an ancestral Assiniboine River probably followed much the same route in preglacial times as the Assiniboine does today. Wisconsinan ice dammed the northward flowing rivers of the province resulting in the formation of lakes, the largest of which, Lake Agassiz, once covered much of Manitoba, as well as extensive areas in northwest Ontario and parts of North Dakota and Minnesota (Teller et al. 1983). To the west, similar large lakes covered parts of Saskatchewan and Alberta (Christiansen 1979) while smaller lakes covered parts of southwestern Manitoba, such as Glacial Lake Souris in the extreme southwest corner of the province. Deposition of fine-grained sediments (glaciolacustrine deposits) on the floors of these lakes resulted in extremely flat topography once the lakes disappeared.

When it was at its highest level, the Herman Stage (Johnston 1946), Lake Agassiz extended far into the Assiniboine Embayment and left behind poorly defined shorelines in the eastern part of Brandon; for example, south of the Assiniboine, one beach runs close to Lorne Avenue between First and Twelfth Streets, and north of the river the "highest shore of Lake Agassiz passes from Brandon east and north-east by Chater and Douglas" (Upham 1896:369). At this time, perhaps twelve thousand to thirteen thousand years ago, meltwater from Saskatchewan entered Lake Agassiz in the area that is now eastern Brandon, depositing a delta which has its apex within the city and fans out to the east. Coarse-grained gravels and sands deposited in the delta have been excavated for construction materials in a series of pits in the eastern part of the city, one of which is now the city dump.

Thus, although the present Assiniboine River has contributed to the excavation of the Assiniboine Valley in which it flows, the valley was cut at the end of the Ice Age by the vast quantities of water flowing from glacial lakes in Saskatchewan to Lake Agassiz. Since then the quantity of water in the valley has decreased so that the Assiniboine River has merely added the details — including the deposition of alluvial sediments on the valley floor — to a form that already existed.

The soils responsible for the rich agriculture in southwestern Manitoba have developed on parent materials of glacial, glacio-fluvial, glaciolacustrine and alluvial origin set down during and after the Wisconsinan Glacial Period. They have developed in an area which experiences a climate with relatively low rainfall and high summer temperatures resulting in a natural vegetation of grassland and aspen parkland. This combination of parent materials, climate and natural vegetation resulted in what are called "Chernozemic Black Soils" (Mills 1984). It is these soils which account for the prosperity of southwestern Manitoba and Brandon, its regional centre. The city is a product of the agricultural region and cannot be considered separately from it.

The theme of the interrelationship between the city and its region is considered in several of the chapters. In Chapter Two the relationships between Brandon and the Assiniboine River are considered because, although the city is a product of the "railway era" and is in some respects a "railroad town," the Assiniboine has played, and continues to play, a major role both in its history and geography.

The climate of southwestern Manitoba is extreme with long, excessively cold winters and short hot summers, and is characterized by relatively low amounts of precipitation. Although this continental climate has contributed to the development of the fertile soils of the region and thus to the region's and the city's prosperity, it also has negative impacts. Large amounts of energy have to be used to overcome the prolonged cold and many activities are excluded or impaired by the harshness of the climate. Moreover, the traditional seasons of temperate regions are greatly modified in southwestern Manitoba. Because the influence of the climate on day-to-day life in Brandon is so profound, its characteristics and some of its more extreme effects are discussed in Chapter Three.

Brandon's total area is now (1987) 4,656 hectares — somewhat large for its population of 38,427 (1984) — a rapid expansion from the small nucleus located along the CPR line south of the Assiniboine River in 1882. The development of this expansion is described and analysed in Chapter Four, while the present population structure and social areas of the city are described in Chapter Five.

The commercial relationship between Brandon and its region is emphasized in Chapter Six, which deals with the city's manufacturing sector which, although not large, is of considerable importance. As is the case in most North American cities, the "downtown area" (Chapter Seven) contrasts and competes with developments on the "urban fringe" (Chapter Eight). Developments in both locations and in the intervening areas are sufficiently volatile that there is a need for urban and regional planning, the background and status of which are detailed in Chapter Nine. Chapter Ten assesses Brandon's role as a service centre within the southwestern Manitoba region, and finally in Chapter Eleven the editors give their opinions on the outlook for the city.

## References

Bird, J.B. 1972. *The Natural Landscapes of Canada.* Toronto: Wiley.

Christiansen, E.A. 1979. The Wisconsinan deglaciation of Southern Saskatchewan and adjacent areas. *Canadian Journal of Earth Sciences* 16(4):913-38.

Elson, J.A. 1959. Surface geology, Brandon — map 1067A. In E.C. Halstead, *Ground-water Resources of the Brandon Map-Area, Manitoba.* Ottawa: Geological Survey of Canada, Memoir 300.

Johnston, W.A. 1946. *Glacial Lake Agassiz, with Special Reference to the Mode of Deformation of the Beaches.* Ottawa: Geological Survey of Canada, Bulletin 7.

McNeil, D.H. 1987. Overview of Cretaceous stratigraphy in the Manitoba Escarpment at Pembina and Riding Mountain, Manitoba. In D.H. McNeil, ed., *Stratigraphy of Albian to Camparian Rocks in the Manitoba Escarpment, Pembina*

*Mountain to Riding Mountain.* Saskatoon: Geological Association of Canada, Guidebook Field Trip No. 2.

Mills, G.F. 1984. Soils in Manitoba. In J.T. Teller, ed., *Natural Heritage of Manitoba: Legacy of the Ice Age.* Winnipeg: Manitoba Museum of Man and Nature.

Teller, J.T. 1984. The Ice Age and its legacy. In J.T Teller, ed., *Natural Heritage of Manitoba: Legacy of the Ice Age.* Winnipeg: Manitoba Museum of Man and Nature.

Teller, J.T., L.H. Thorliefson, L.A. Dredge, H.C. Hobbs and B.T. Schreiner. 1983. Maximum extent and major features of Lake Agassiz. In J.T. Teller and L. Clayton, eds., *Glacial Lake Agassiz.* St. Johns: Geological Association of Canada, Special Paper 26.

Upham, W. 1896. *The Glacial Lake Agassiz.* Washington: United States Geological Survey, Monograph 25.

Welsted, J. and H.R. Young. 1980. Geology and origin of the Brandon Hills, Southwest Manitoba. *Canadian Journal of Earth Sciences* 17(7):942-51.

# TWO
# Brandon and the Assiniboine River

*John Welsted*

Although Brandon is clearly a railroad city, it owes much to its location within the valley of the Assiniboine — the major waterway of southwestern Manitoba. Both the shape and form of the Assiniboine Valley and the flow characteristics of the river have been major factors in the growth, development and life of the city.

## The Valley

The Assiniboine Valley is a glacial spillway that used to carry water from glacial lakes in Saskatchewan to Glacial Lake Agassiz. West of the city, it has a typical spillway shape with steep sides and a flat floor a kilometre or so across, but it loses this form where the Assiniboine Delta begins. At Brandon the north side of the spillway is usually steeper than the south side on which most of the settlement is located (Figure 2:1). Both north and south slopes are terraced, the terraces probably relating to the various base levels assumed by the Assiniboine as Lake Agassiz emptied and decreased in level. However, no attempt has yet been made to relate these terraces to specific lake levels.

Soils on the floor of the Assiniboine Valley are a product of the river, being predominantly alluvial materials of silty clay texture. The texture does, however, vary from one site to another and in some locations materials are still deposited during spring floods, the most recent severe flood being in 1979. "Since these materials are of recent origin, soil profile formation is, for the most part, poorly developed. Under woodland vegetation conditions where the alluvial material is moderately well-drained, profile development is the most pronounced. Here a weak, greyish 'A2' layer and a slight [*sic*] textured 'B' horizon can sometimes be observed. At the same time upon high well drained banks the soil profile is also partially developed" (City of Brandon 1974). Where the soils are well-drained, as for example in the area of the Experimental Farm and in parts of "the flats" north of the CPR line, they provide good

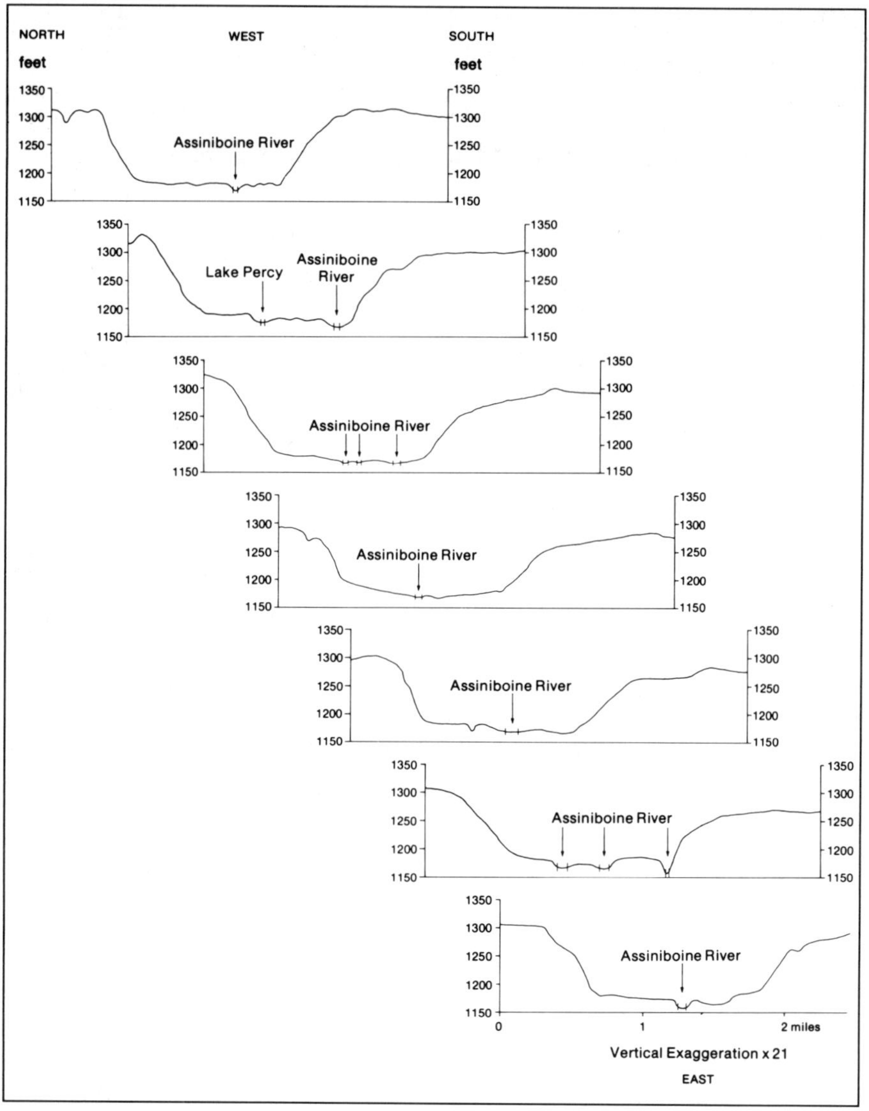


FIGURE 2:1
*Cross profiles of the Assiniboine Valley at Brandon*

agricultural and horticultural land. Poorly drained abandoned meanders, though, are too wet for cropping and can usually be picked out in summer by their covering of grass as opposed to crops.

## The River

The Assiniboine rises in Nut Mountain in eastern Saskatchewan and drains an area of 92,204 square kilometres. It enters Manitoba west of Roblin and near St. Lazare is joined by the Qu'Appelle River (Figure 2:2). The other major right bank tributaries are the Souris, which joins at Treesbank below Brandon, and the Cypress River which joins downstream from Spruce Woods Provincial Park. Left bank tributaries draining from Duck Mountain and Riding Mountain are short, the major ones being the Shell, the Birdtail Creek, Oak River and the Little Saskatchewan River — all of which join above Brandon. Epinette Creek, a left bank tributary which joins below Brandon, has its headwaters at Sewell Lake. It follows the route of a former distributary across the delta — the path followed about eleven thousand years ago by a large proportion of the Assiniboine's flow on its way to Lake Agassiz. Diversion of the bulk of the flow of the Assiniboine to its present course along the southern edge of the delta is probably associated with rebound from ice loading causing land in the north to rise with respect to that further south, rebound having been greatest in the north where the ice was thickest. This process is also responsible for the general southward slope of the once horizontal Agassiz beaches (Johnston 1946).

Above Brandon the Assiniboine is a freely meandering river, wandering its way across the floor of the glacial spillway. In the stretch between Shellmouth and Brandon, meander cutoffs and oxbow lakes are abundant, eighteen meanders having been abandoned in the past one hundred years. In the short stretch between the junction of the Little Saskatchewan and the western boundary of the city, the river is less sinuous, mainly due to the abandonment of a series of meanders on the northern bank. These abandoned meanders are best seen during the spring when they are occupied by water. Within the city limits, two abandoned stretches of the river are identified by name — Lake Percy north of the main course in the area of the Experimental Farm (Agriculture Canada Research Station) and the Snye to the south (Figure 2:3). Although the latter has now almost disappeared, in the early days of the city riverboats used it, residents swam in it and a bridge was needed to cross it. It was eventually blocked off in 1925 as a flood control measure, but it is still a wet area and can be clearly identified on air photographs and on detailed maps by the absence of settlement along it.

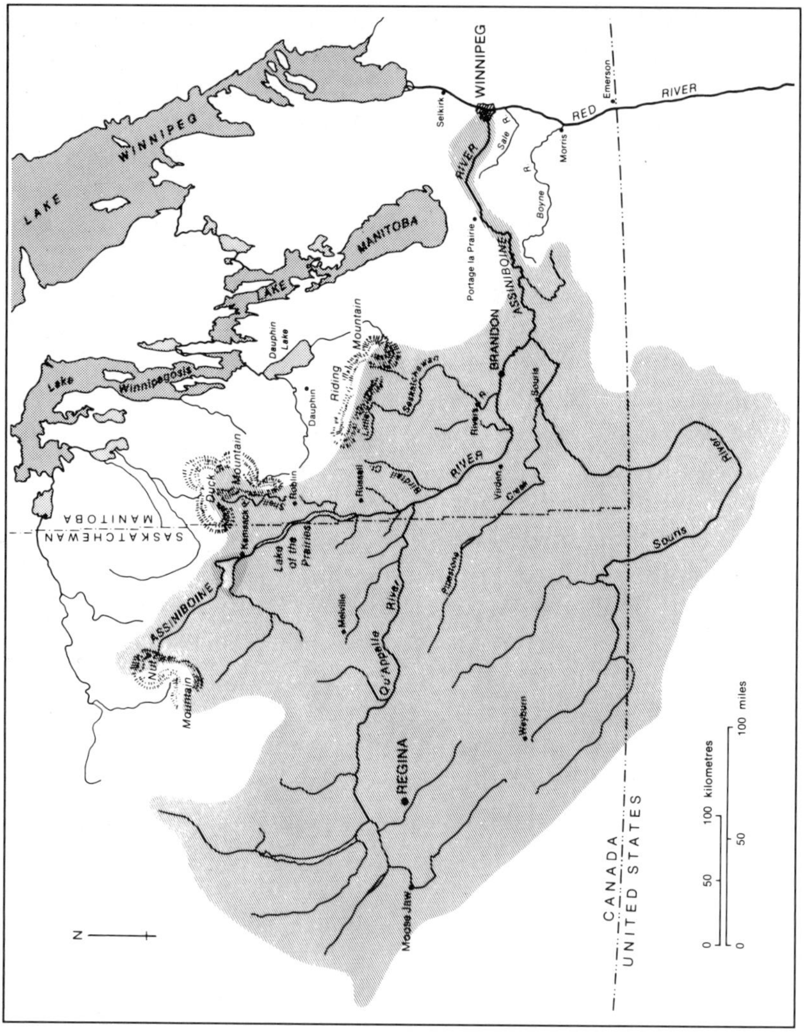


FIGURE 2:2
*The Assiniboine Basin*

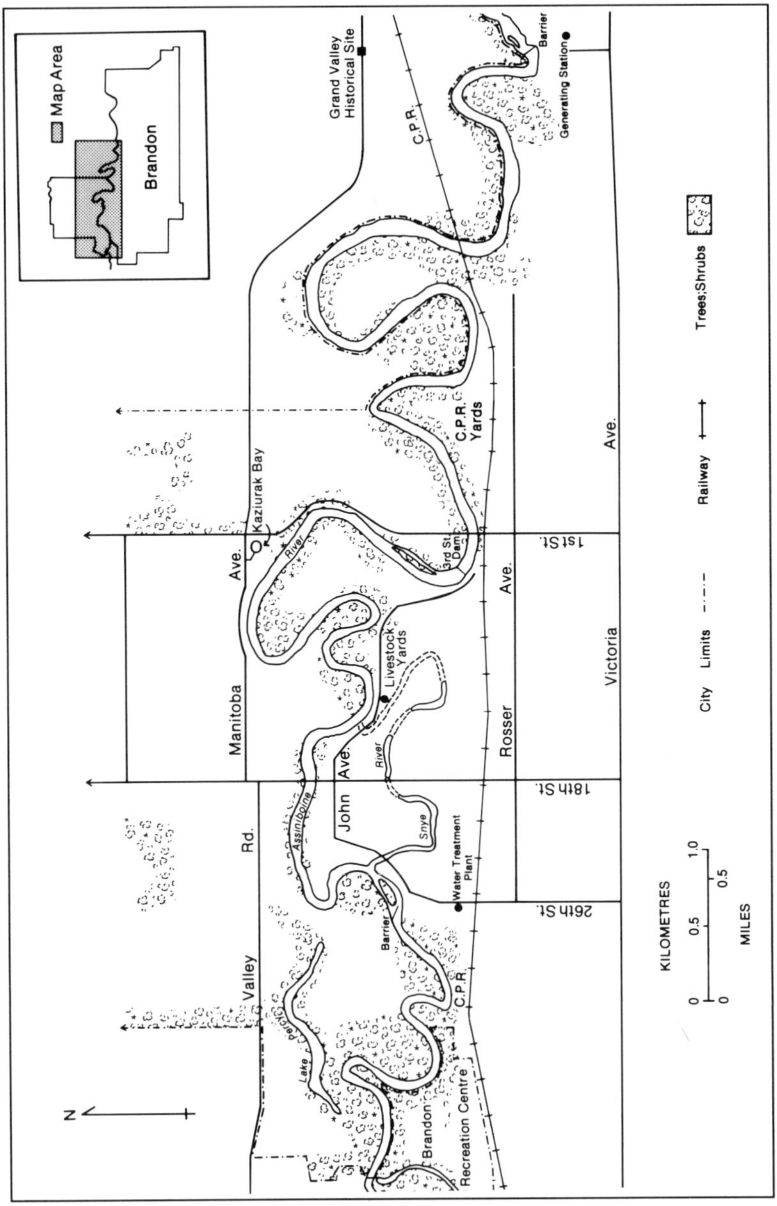


FIGURE 2:3

*Location of natural and human-made features in the Assiniboine Valley at Brandon*

Below the city, the Assiniboine continues to meander but, as it is incised into the sands and gravels of the delta, it is less free to change positions. Nevertheless, some very obvious abandoned meanders can be found; for example, Marsh's Lake in Spruce Woods Provincial Park. A few kilometres downstream from Brandon, at Currie's Landing, a series of rapids may represent a knickpoint produced as the Assiniboine adjusted to a falling local base level — now the Red River, whereas in the past it was the higher level of Lake Agassiz.

As might be expected from its meandering form, the gradient of the river is gentle. It falls only one metre from Curran Park, just west of the city boundary, to First Street Bridge, a channel length of 10.5 kilometres, but a straight line distance of only four kilometres. Although the flow is not normally rapid, averaging only one kilometre per hour in the fall, it increases to three kilometres per hour in the spring when discharge is normally greatest.

The flow regime is typical of prairie rivers with maximum discharges usually being recorded in May which has a mean flow of 106 cubic metres per second (cms) with the next highest month being April (mean, 90 cms). Between 1906 and 1973 discharge was recorded at First Street Bridge, but the gauging station has since been moved upstream near the Trans-Canada Highway crossing. The values for discharge given above are for the new location; equivalent values for the First Street location are slightly different. Maximum discharges are associated with snowmelt, the highest maximum instantaneous discharge (the peak flow) being 262 cms recorded at 16.38 hours on 1 May 1975. Values were probably higher than this on 18 April 1976 when a very high daily value was recorded but there is no record of the maximum instantaneous discharge for that day. Minimum discharges occur in winter with February being the lowest month (mean, 16.4 cms at the new gauging station, 9.43 cms at First Street). The lowest minimum daily discharge ever recorded was only 0.566 cms on 11 December 1936. In recent years the regime has been affected by a flood control dam at Shellmouth, in operation since 1969, which has reduced discharge in spring and increased it during the summer.

Within the city, the river is from forty-five to seventy-five metres wide, enough to form a major barrier to north-south travel. At normal August levels it averages about 1.2 metres in depth, but water depths are influenced by the dam at Shellmouth and by a weir at Third Street (City of Brandon 1974). Cross channel profiles are typical of meandering streams, with even profiles along the straight stretches and asymmetrical profiles at the apex of bends, the deep points being located on the outside (Figure 2:4).

Freeze-up of the river normally occurs in mid-November but has been as early as 20 October and as late as 8 December; break-up time is typically mid-April, but has occurred from 27 March to 1 May (City of Brandon 1974).

## Transport

Although river transport on the Assiniboine is no longer significant, it was an important element in the settlement of the Assiniboine Valley. Between 1878 and 1885 a series of steamers carried settlers and goods between Winnipeg and Fort Ellice at the junction of the Assiniboine and the Qu'Appelle (Figure 2:2):

> Three ships plied the Assiniboine in 1878. Most of the traffic that year was between Winnipeg and Portage La Prairie but, with the increase in settlement of the Little Saskatchewan, the *Manitoba* was chartered for the foot of the Rapids (by Currie's Landing), bringing in large numbers of immigrants. A regular service was provided in 1879 by the *Marquette* and the *Alpha*, flat bottomed stern wheelers drawing 18-26 inches of water. That year, too, both vessels ventured beyond the Rapids to Fort Ellice, which was regarded thereafter as the normal head of navigation, although at high water vessels occasionally went as far as Fort Pelly [north of Kamsack in Saskatchewan] . . . The "Navy" was busiest in 1881 when the *Marquette* and the *Alpha* were joined by the *North-West*, the *City of Winnipeg* and the *Manitoba* in carrying both freight and passsengers to the Rapid City landing (Currie's) and Grand Valley. . . . In 1882, with a railway parallelling the river, the traffic did not warrant use of additional vessels. During the greater part of the season, which closed earlier than usual because of lower water, the *Marquette* went only as far as Brandon. The upper portion of the river was served by the *Alpha*, which made one trip to Fort Pelly, but for the balance of the season ran only to Strathallan at the confluence of the Qu'Appelle and Assiniboine rivers. (Tyman 1972:32)

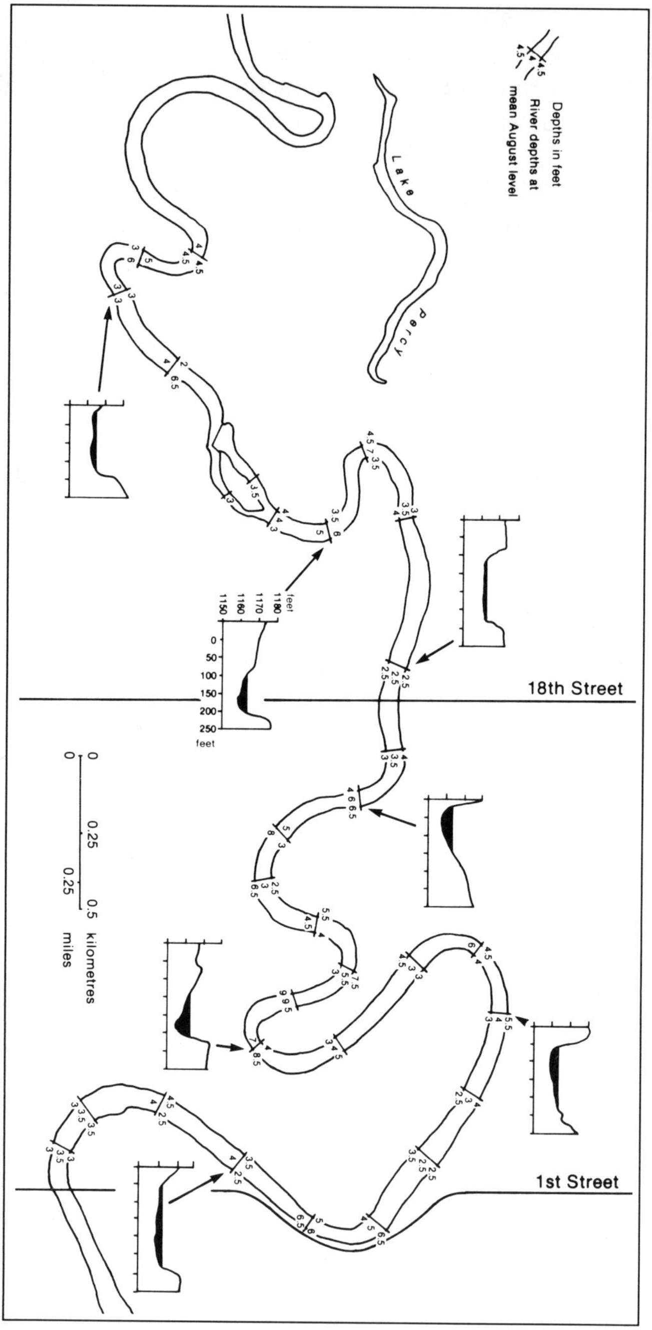


FIGURE 2:4
*Cross channel profiles of the Assiniboine River at Brandon*

The *Alpha* continued to travel the river until 1885, but on 27 April 1885 it ran aground "while attempting to cross a peninsula in flood waters" (Brown 1981:66). The Assiniboine has been of no significance as a navigation route since.

However, the river was used to transport logs in the early part of this century (Plate 2:1). "Spring time witnessed log-drives that began in Northern Manitoba forest areas and ended at Brandon," where logs were assembled in the Snye. "The 1910 drive — five million board feet, perhaps the largest ever brought down the Assiniboine — began on March 31st, but not until mid-June did the one hundred and twenty-five peavy-wielders (many among them members of Winter tree-cutting teams) reach this city" (Barker 1977:127-28).

**Bridges**

The Assiniboine is a significant barrier to north-south travel in the Brandon area. However, there was soon a need to travel between Brandon, developing on the gently sloping south side of the spillway, and the small villages and towns to the north. Initially, the river was crossed by a ferry near Third Street (Plate 2:2) but two bridging points a mile apart were used from an early date, one at First Street and the other at Eighteenth Street. *The Commercial* (10/10/1882) reported that two bridges had been completed at these locations. They must have been completed during the summer of 1882 because in his recollections of the "great flood" of 1882 White writes:

> At that time there had not been any bridges erected, . . . There was no obstruction to dam the water and it just swept over everything. The only means of crossing the river at this time was a ferry near the site of the Western Canada Flour Mills [located at the junction of Fourth and Assiniboine Streets], and the ford at Grand Valley, but both ferry and ford were soon out of business. (*The Brandon Daily Sun* 18/06/1927)

The bridge at First Street was first constructed by J.E. Woodworth at a cost of $28,000 while that at Eighteenth Street is estimated to have cost $30,000, the greater cost of the latter being due to the construction of numerous approaches to the bridge proper, one of which had to cross the Snye. Originally both were toll bridges, but they were eventually purchased by the city and thrown open to the

PLATE 2:1

*Logs stored on the Assiniboine during Hanbury's "big drive" of 1910 at the peak of the sawmilling industry in Brandon. In that year 125 drivers brought five million board feet of timber down the Assiniboine. Logs were cut in what is now Riding Mountain National Park and floated down the Little Saskatchewan River and Birdtail Creek. As their usual storage area, the Snye River, was filled, some were stored on the Assiniboine. (Lawrence Stuckey: personal collection)*

PLATE 2:2

*Ferry crossing the Assiniboine at about Third Street in 1882. The Assiniboine is a substantial barrier to north-south travel, so at an early stage in Brandon's history bridges were constructed at First Street and Eighteenth Street and the ferry was abandoned. (Lawrence Stuckey: personal collection)* public, although not without considerable debate, as recorded by *The Commercial* (24/07/1883):

> A strong meeting was held against purchasing these miserable structures which are admitted by all not to be worth the price of the wood in them. The cost of repairing will be greater in four years than the cost of the bridges. The feeling of the public is that a good iron bridge should be erected in the centre of the city.

The original bridge at Eighteenth Street was a swing bridge to allow high river boats to pass. "It had a very extraordinary swing, the centre span revolved upon a pivot and trembled and shook so much as vehicles passed over it that horses and oxen became very frightened and much excitement ensued" (*The Brandon Daily Sun* 18/06/1927). This bridge and the tolls levied for it were so controversial that in 1885 a group of farmers from north of the river dumped the tollhouse into the river causing the bridge lessee, E. Hoskin, to withdraw from the bridge business and to leave the city (Barker 1977:10).

The First and Eighteenth Street crossings have persisted, with necessary upgrading, to the present. A bridge at First Street, crossing both the river and the railway tracks, was built in 1909 by the city and the CPR. It was replaced in 1972 by a modern structure, costing approximately $1.5 million, after much wrangling between the city, the province, and the CPR as to who was responsible for its upkeep. The present structure at Eighteenth Street, the David Thompson Bridge, was completed in 1960, replacing the old bridge built in 1907.

Another bridge existed for a time at Seventeenth Street East. This small bridge, built in 1921, was eventually removed, presumably because it was not much used. However, there may soon be another bridge in this general area when the long-awaited bypass to the industrial park in the eastern part of the city is constructed. To the west, the nearest bridge is approximately eight kilometres upstream where the Trans-Canada Highway crosses the Assiniboine near Kemnay.

The first written reference to a railway bridge across the Assiniboine was by William Robertson, captain of the *Alpha*. As his ship left Brandon for Winnipeg on Wednesday, 20 June 1883, he made notes on one of his navigational maps, just below Brandon, recording that "We have been delayed here at the CPR swing bridge. Some problem exists with the mechanism that allows the bridge to open and let us pass. This bridge will likely be replaced once shipping on the upper Assiniboine ends" (Province of Manitoba 1985).

**Water Supply**

The city of Brandon gets its water from the Assiniboine. At an early date the need for a reliable water supply system was recognized and on 12 September 1893 the first waterworks system went into operation with the corporation announcing a schedule of fees that was designed to encourage the use of water. At that time there was an annual eight-dollar charge for a dwelling place containing up to six rooms, a twelve-dollar fee for a six-room boarding house, and a hotel levy of two dollars for each room. These charges were presumably for unlimited use — metres were not used in the city until 1938. Prices have increased, of course, but water is still cheap and it is still priced to encourage use. The present rates (1987) are $1.74 per 100 cubic feet for the first 5,000 cubic feet (141.5 cubic metres) per quarter (domestic rate); $1.34 for the next 95,000 cubic feet (2,688 cubic metres) per quarter (intermediate rate); and $1.00 for anything over 100,000 cubic feet (2,830 cubic metres) per quarter (wholesale rate).

In the past, seasonal fluctuations in the volume of the Assiniboine, with minimum flows during the winter of under ten cubic feet (0.3 cubic metres) per second, caused water supply problems. However, Shellmouth Dam has improved the reliability of flow at Brandon; it allows for a regular flow of three hundred cubic feet (8.5 cubic metres) per second throughout the year. A weir built across the river at Third Street ensures an adequate water level during most years, but low water levels during the summer of 1988 brought suggestions that the level of the weir be increased (*Brandon Sun* 05/08/1988).

The treatment plant, built in 1893, was located at the foot of Twenty-Sixth Street and has remained there ever since, although it has had additions and improvements. The present facility is actually three plants located together: the first, with a capacity of three million gallons (13,638,000 litres) per day, was in operation by 1948; the second, also with a three-million-gallon per day capacity, was in operation by 1958; the most recent addition, built in 1978, can treat six million gallons (27,276,000 litres) per day for a total capacity of twelve million gallons (54,552,000 litres) per day. Consumption is well below this figure; for example, average consumption was 3,676,155 gallons (16,711,800 litres) per day in 1982, 3,667,758 gallons (16,673,627 litres) per day in 1983, and 3,715,576 gallons (16,891,000 litres) per day in 1984. However, these averages hide great variations in daily use, the extremes for 1984 being a low of 2,379,000 gallons (10,814,934 litres) on 25 January and a high of 8,376,000 gallons (38,077,296 litres) on 25 July. Summer consumption is considerably higher than winter consumption, mainly because of garden use of water. In some dry summers the demand has been so great that residents have been asked to reduce the watering of lawns (*Brandon Sun* 09/08/88). Virtually all homes within the city limits are now served by city water but many use private wells for lawn watering.

Water is stored at three locations. At the water treatment plant there is a 450,000 gallon (2,045,700 litres) underground storage facility under the filter system plus an additional 400,000 gallon (1,818,400 litres) storage area. Water stored in these areas is used for backwash purposes and to meet emergencies and/or peak demand conditions. Second, there is an elevated tank (water tower) located in Rideau Park in the east end of the city capable of holding 675,000 gallons (3,068,558 litres) which feeds water into the distribution system by gravity. Finally, an above ground storage reservoir, with a four-million-gallon (18,184,000 litres) capacity, is located at the junction of Tenth Street and Brandon Avenue. Treated water from this storage unit is forced into the distribution system by means of booster pumps.

The water treatment plant is not the only place at which water is extracted from the river. Since 1967, Manitoba Hydro has operated a thermal electric generating station on the south bank of the river in the eastern part of the city (Figure 2:3). It is a peaking station and only operates for a short time during the winter when flow through the major water power stations in northern Manitoba is depleted. During the winter of 1986-87 the station was in operation from November to mid-March. When it is operating it uses water from the river at a rate of two hundred cubic feet (5.66 cubic metres) per second. This is a nonconsumptive use of water — for cooling — virtually all of it being returned to the river very close to where it is taken out, but at a much higher temperature — 28°C. At the other end of the city, water is pumped from the river to irrigate the city's golf course, and at other places, particularly in the Kasiurak Bay area, householders pump water from the river for use on gardens (Figure 2:3). The Mental Hospital, on Brandon's north hill, pumped water from the river at First Street until a few years ago.

The Assiniboine is the primary but not the only source of water in Brandon. Groundwater is used by a few high-water demanding industries in the city such as Manitoba Pool Elevators, Canadian Occidental Petroleum Limited (which has a well one hundred feet deep) and Simplot Chemicals, which obtains water from three two-hundred-foot deep wells. The water is hard but it is suitable for the various processes involved. In addition, the city obtains water from a well to supply the swimming pool at Curran Park, and a few rural-type residences within the city limits still rely on groundwater.

In the past the Assiniboine was a source of ice, for use in both the city and other parts of western Canada:

> The frozen harvest period usually extended from mid-January to early March during which time the almost transparent ice measured between twenty-five and forty inches thick, producing solid blocks of one thousand pounds each. The field (stretching from Ninth to Eighteenth Streets) was used annually by both outside and local firms. The Arctic Ice Company of Winnipeg looked after Canadian Pacific Railway western requirements; Lane and Elviss, together with R.E. Rice, packed for Brandon district needs. (Barker 1977:128)

## Waste Disposal

Traditionally rivers have been used for sewage disposal, the Assiniboine being no exception. Before 1964, untreated sewage was deposited in the river at two or three locations, but after this date the city was required to develop and install a sewage treatment system. The present system, which is located in the east end of the city, consists of two phases. The first is an extended treatment plant located in the northwest quarter of Section 17, Township 10, Range 18 on the south side of the river. It is a four-million-gallon (18,184,000 litres) aeration basin used for effluent during the winter months when lagoon treatment is ineffective. Clarified sewage is pumped from here under the Assiniboine River to lagoons located north of the river. The lagoons provide second stage treatment after which the treated effluent is discharged into the river.

The system has a capacity of six million gallons (27,276,000 litres) per day and can handle the city's sewage during the summer, but problems occur in the winter when the lagoons are frozen and there is heavy discharge from Ayerst Organics in the eastern part of the city. An estimated $15.6 million expansion to the system has been studied with construction being spread over a three-year period for completion in 1991.

Some of the major industrial plants in the city have their own waste disposal systems; for example, Canadian Occidental Petroleum Limited stopped discharging waste into the river in 1982 and now deposits it into a landfill site at the plant. Also, Simplot Chemicals Limited disposes of some of its waste by using it for irrigation of nearby fields, some of which it owns and some of which are owned by neighbouring farmers. This process, which has been termed "fertigation" is normally practiced only between 15 May and 30 September, but under favourable weather conditions the process can start a month earlier. The firm is permitted to discharge waste into the river during the period 15 May (15 April when the weather is favourable) to 15 September although it has to be within limits prescribed by the Manitoba Clean Environment Commission.

## Recreation

Another traditional use of rivers is for recreation, but in many areas with increasing in-wash of herbicides and pesticides from agricultural land, with obstructions to the natural flow, and with increasing urbanization and industrialization of banks and floodplains, recreation has become less attractive. Although the Assiniboine is not a major focus of recreational activity, it is used for a number of purposes, some of which were more important in the past than they are now.

Although swimming in the river was popular in the past, it is no longer common even though "it usually meets standards for whole body contact recreation" (Province of Manitoba 1976:1-14). During the early 1930s there was a popular bathing beach on the banks of the Assiniboine at the foot of Fifteenth Street. At that time, concession stands operated all summer long and gas lights made nighttime swimming possible.

Fishing in the river has been popular from the early days. The Assiniboine provides a variety of habitats for more than thirty species of fish including walleye, sauger, perch, northern pike, mooneye, rock bass, catfish, bullhead, white sucker, silver redhorse, shorthead sucker, quill back sucker, barbot, carp, chestnut lamprey, and numerous minnows (Glenn 1977). The preferred locations for fishing are at the Third Street dam and near the weir at Manitoba Hydro's steam generating plant. An indication of the popularity of the river is given by the fact that a spot check in 1984 revealed 250 people fishing between the Third Street dam and the hydro plant, and a fishing derby in May 1984 had 61 of 170 competitors fishing the Assiniboine below the Third Street dam. Some people still ice fish in the winter.

Boating is now limited, although a popular canoe trip for Brandon residents is along the Assiniboine from the Trans-Canada Highway crossing, west of the city, to the David Thompson Bridge. Downstream from Brandon an interesting canoe trip can be taken to Spruce Woods Provincial Park. The province publishes a descriptive canoe-route map for this journey with added commentary written in 1883 by Captain Robertson of the *Alpha* (Province of Manitoba 1985).

Recently, water skiing has been practiced on the river. Although part of a large meander loop, the stretch of the river between Manitoba Avenue and First Street is relatively straight (Figure 2:3). It also is sheltered which limits the size of waves; the north bank provides a good viewpoint for spectators; and, although the river is narrow, it was wide enough for the Western Canada Water Ski Championships in 1982. In the winter, snowmobiling on the river, one of the few places within the city where snowmobiles are allowed, is popular.

In addition to these recreational activities which are directly water related, several recreational areas take advantage of the floodplain. On the north shore, Curran Park has a swimming pool, several baseball diamonds, and facilities for camping and picnicking. South of the river, much of the city's golf course is located within the floodplain. Further east, Queen Elizabeth Park includes tennis courts and picnic facilities and the Westbran Work Activity Project operates a ferry across the river to serve Curran Park during the summer months. This project is a skill training programme with the secondary objective of developing the Assiniboine floodplain according to the needs of the city and region. As a result playing fields have been laid out and an activity trail has been developed. East of Eighteenth Street, Eleanor Kidd Park offers picnic facilities as does Dinsdale Park west of First Street. East of First Street a soccer field complex has been laid out and should be in use in 1989. Eventually, if plans are carried to fruition, a recreational corridor will extend from the golf course in the west to beyond First Street in the east. Unfortunately, the corridor will be broken between Eighteenth Street and First Street, by preexisting and long-established industrial and commercial development south of a flood control dike.

## Problems Associated with the River

Although Brandon's location within the Assiniboine Valley and its proximity to the river have been positive factors in the development of the city, they have resulted in problems as well. Probably the most persistent of these has been flooding, but bank erosion, pollution, and the question of jurisdiction over the river are problems which have caused trouble and which could become more severe in the future.

### *Flooding*

In common with other meandering rivers, the Assiniboine is subject to periodic flooding, defined as a situation in which the river "rises to overflow land which is not normally submerged" (Ward 1978:5). Although normally the flow of the Assiniboine is confined to a well-defined channel which meanders through Brandon, at frequent, but irregular, intervals in the past it has spread across much of its floodplain (Dechka 1981).

Flooding by the Assiniboine has influenced the growth of the city since the earliest times. The first settlement in the immediate Brandon area was at Grand Valley, north of the river three kilometres east of the present Central Business District (Figure 2:3). When it was decided that the CPR would pass through Portage La Prairie rather than taking a more northerly route, it was assumed that Grand Valley would be the location of a rail divisional point, and by 1881 the population of the settlement had grown to four hundred. However, landowners in the area asked for more than the CPR was prepared to pay and Colonel Rosser, the engineer in charge at the time, decided on a higher, well-drained site on the south side of the river three kilometres upstream. This proved to be a wise decision because in 1881 Grand Valley was submerged beneath two metres of water and was flooded again during the spring of 1882. People either abandoned their homes and businesses or moved them to the new settlement of Brandon. All that remains of the town of Grand Valley is a cairn and one building indicating its former site. Settlement in the new city of Brandon spread from Pacific Avenue, immediately south of the railway tracks, southward up the gentle slopes of the spillway and northward onto the Assiniboine floodplain into the area of "the flats" where flat terrain and fertile soil were an attraction despite the risk of flooding.

Flooding has occurred in twenty-nine of the seventy-nine years for which records are available, with major floods, defined as those which have resulted in property damage, occurring in seventeen of these years (Dechka 1981). Flooding usually resulted when peak river levels at First Street exceeded 1,170 feet (356.6 metres) above datum, a situation which has occurred in both spring and summer seasons. Spring floods, which usually result from a combination of high antecedent soil moisture conditions, rapid snow melt and heavy rainfall, account for twenty-six of the flood years (Plate 2:3). As an example, during the spring of 1923 two peaks were recorded, the first on 20 April and a higher one on 7 May. They resulted from ice jams and heavy snow and rainfall. High antecedent soil moisture conditions meant that water from melting snow was unable to infiltrate and consequently ran off into already filled drainage

PLATE 2:3

*Spring floods have been a recurring feature of Brandon's history. This photograph, which was taken in 1922 looking northwards across "the flats" (now called Assiniboine Gardens) to the North Hill, shows the extensive flooding of residential property. Brandon Mental Hospital can be seen on the skyline. (Public Archives of Manitoba: Water Survey Collection G1132)*

ditches. During this flood, the highest that has been well documented in Brandon's history, water entered homes through windows and foundations collapsed. Other damage included the loss of lumber and the washing away of small shacks and poultry buildings on "the flats" (*The Brandon Daily Sun* 07/05/1923).

The most recent summer flood occurred on 9 July 1954 as a result of heavy rainfall. At this time the Valley Road was covered by water and families were forced to evacuate homes as water filled basements and covered floors. Also affected were market gardens on "the flats," the area now occupied by Queen Elizabeth Park, and the first and second holes of the golf course at the Brandon Recreation Centre (Figure 2:3). Eighteenth Street was damaged and the destruction of a one-hundred-metre section of a dike resulted in flooding behind it. Damage was estimated to total $65,000 (*The Brandon Daily Sun* 16/05/1955).

The spread of flood waters is influenced by structures which have been built on the floodplain. Both First Street and Eighteenth Street cross the floodplain from north to south, blocking the movement of flood waters. These roads are now high enough to provide access to the city during all except the highest floods, the most recent exception being the 1976 flood which over-topped Eighteenth Street; First Street was protected by sand bagging at that time. The Snye River was blocked at its upstream end in 1925, resulting in a reduced flood area during minor floods but during major floods the dike is over-topped and the former channel occupied. In 1954, the city built a dike from Third Street to Twenty-Sixth Street with a road, known as John Avenue, at its crest (Figure 2:3). This dike has been continually raised over the years and is now at an elevation of 1,182 feet (360.3 metres) above datum. It prevents flooding of the area to the south of it but has the adverse effect of constricting the floodway (the area occupied by flood waters), causing water to back up to the west.

Shellmouth Dam, a major structure on the Assiniboine River 210 kilometres northwest of Brandon, has affected the extent of flooding within the city in recent years. It is part of a flood control project for the Assiniboine and Red Rivers which also includes a diversion channel from the Assiniboine to Lake Manitoba west of Portage La Prairie, and the Red River Floodway around Winnipeg (Mudry et al. 1981). Although it did not prevent flooding at Brandon in 1976, the worst flood in more than fifty years, without Shellmouth the flood would probably have exceeded that of 1923.

In recent years the emphasis in flood mitigation in Canada has turned from structural to nonstructural measures, evidence of which can be found in the Canada-Manitoba Flood Damage Reduction Program which aims to map the distribution of flood risk areas and to discourage people from locating in them. As part of the program, a flood risk map for Brandon has been produced (1981) at a scale of 1:25,000. On it are designated a "floodway" and a "floodway fringe," the floodway being the area in which flood waters would be deepest and along which the water would continue to move, and the floodway fringe being an area in which flood water would pond and move sluggishly. Dechka (1981) produced a similar map which differed from the Flood Damage Reduction Program map in that he recognized a considerably larger floodway fringe extending to the 1,190 foot (362.7 metre) contour line whereas the official map used the 360 metre (1,181 foot) contour as the boundary. Although the documentation is limited, evidence exists that the floods of 1882, 1902 and 1904 exceeded the 360-metre level. Certainly the flood of 1923 did and the 1976 flood was almost as high. For this reason, it is thought that the official map seriously underestimates the flood risk at Brandon.

One of the results of the repeated flooding is that there is little residential settlement left on the floodplain although it used to be heavily settled (Plate 2:3). Until recently, the only modern development on the floodplain was at Kasiurak Bay, north of the river near First Street (Figure 2:3) which seems to be at risk even though it is not included in the flood risk area on the Flood Damage Reduction Program map. Perhaps it is omitted because a substantial dike has been built between it and the river. However, this dike is lower than the flood levels of 1882, 1902 and 1904. With this in mind it is amazing that, in 1986, construction of a number of residences started north of Manitoba Avenue on land that is undoubtedly flood prone. Cursory examination of these houses indicates that there is no obvious flood proofing. Another surprising exclusion from the flood risk area is the Grand Valley Historical Site which was most definitely flooded in 1881 and 1882.

There is a lot of commercial development, some old and some new, within the floodway fringe protected by the John Avenue dike south of the river. In April 1982 stockyards in the floodway fringe were destroyed by fire. These livestock yards did not conform to the zoning under the city's 1979 development plan but, because the plan was drawn up after the yards were built, the facility was considered a nonconforming, but permitted, use. According to the plan, if more than 60 percent of a nonconforming business is destroyed by fire, rebuilding is not allowed. However, the owners estimated that only 35-40 percent of the operation was destroyed and city council gave permission for the yards to be rebuilt.

Much of the recent development on the floodplain has been for recreational purposes—parks, activity areas, ornamental ponds, and a golf course. Investment in structures for these facilities is minimal; damage incurred by flooding is not great and, on the whole, it is not deemed necessary to protect them from flooding. However, repeated flooding of the lower half of the golf course has restricted its use to nine holes in the early part of the season. Consequently, a dike has been built with its crest at 1,180 feet (359.7 metres). This would have been sufficient to protect the course against all except the highest floods (i.e., those of 1882, 1902, 1904, 1923 and 1976).

*Bank Erosion*

Meandering rivers erode on the outside of bends and by doing so the meanders gradually change position (Figure 2:5). Erosion of this nature has caused some problems, particularly along McDonald Avenue where one small house has been demolished at the top of a steep river cliff and two others have been threatened over the past ten years. A stepped wall of railway ties has been built to protect the bank at this point. The wall and the reduction in the spring discharge of the river caused by the Shellmouth Dam have insured that the houses are reasonably safe. On the north side of the river, Manitoba Avenue, near its junction with Sixth Street North, now (1988) curves around a meander loop whereas it used to be a straight east-west road. The bank here is protected from erosion by debris derived from the demolition of the Prince Edward Hotel which used to exist in the centre of the city. Other protected stretches include the bend in Curran Park (north bank), John Avenue dike at Fifteenth Street North (south bank), the bend at First Street (east bank), and the CPR line (south bank).

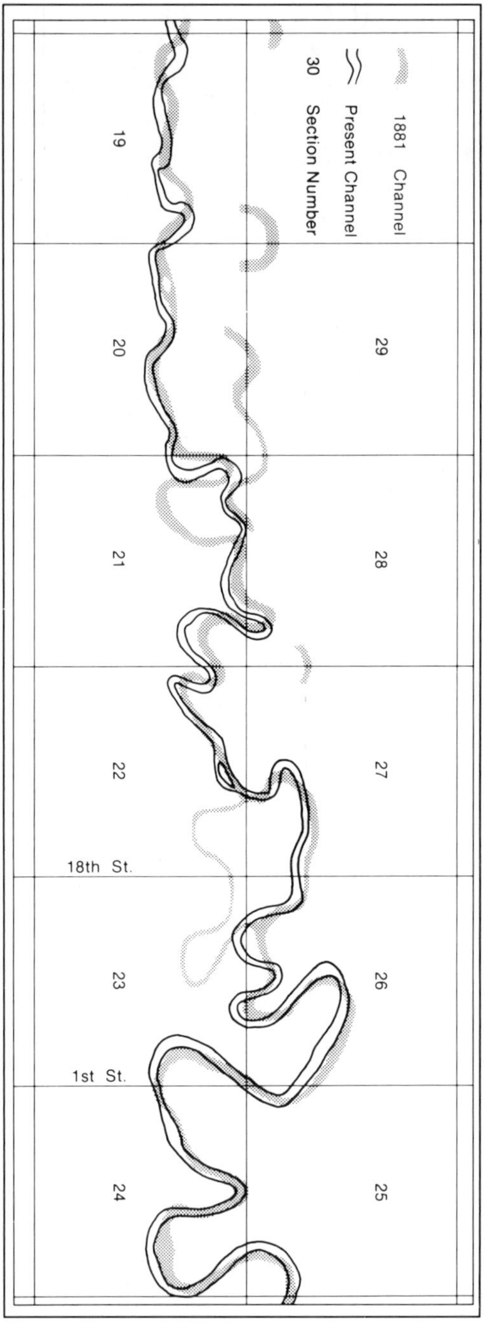


FIGURE 2:5
*Changes in the course of the Assiniboine River between 1881 and 1987*

*Pollution*

Pollution of the river has not been eliminated although environmental control standards are now more stringent. Routine dumping of raw sewage into the river at Brandon and by settlements upstream is forbidden but in October 1977, and again in October 1981, raw sewage was discharged into the river when the pumping plant which sends sewage to the lagoons was unable to cope with the volume. The new sewage treatment plant soon to be started should eliminate this problem. However, approximately 15 percent of the city's sanitary system is still linked to storm sewers and overflow of raw sewage into the river occurs once or twice a year at times of heavy precipitation, most recently on 5 May 1986.

Also, Simplot Chemicals discharges waste into the river — a matter of considerable concern to downstream towns and municipalities, particularly Portage La Prairie with its food-processing industry. As recently as August 1985, Simplot had problems in meeting the standards for discharge of ammonia-nitrogen (90 kilograms per day) set by the Manitoba Clean Environment Commission but now claims to meet the standard 95 percent of the time. Manitoba Hydro has had trouble too; in October 1981, 1,000 gallons (4,546 litres) of heavy fuel oil were spilled into the river. It was eventually burned off, most of it having accumulated in a small tributary creek, but the clean-up operation cost approximately $10,000. Thermal pollution is represented by the return temperature of 28°C of water used by Manitoba Hydro for cooling. This water is harmful to fish within the immediate vicinity. However, specific fish kills, such as that which occurred near the Spruce Woods Provincial Park in July 1976, have not been attributed to this cause.

The main problem for the operators of the water treatment plant is nonpoint source pollution — run off from fields within the Assiniboine Basin. This leads to algae growth in the spring and problems with taste and odour in the domestic water supply. It has proved difficult to predict what chemicals will get into the river and what will happen to them once they are there. Improved treatment starting in the spring of 1985 ensured fewer complaints about taste and odour for three years but the hot, dry weather of the summer of 1988 caused a renewal of the problems with many people resorting to taking drinking water from a spring on the North Hill (*Brandon Sun* 03/08/1988).

*Jurisdiction*

Jurisdiction over the river is not clear. The river runs through Brandon and it is clearly of great relevance to the city, which therefore desires some control over it. However, both to the west and to the east the river forms part of the boundary between the city and the surrounding municipality of Cornwallis, running along the south bank in the west and along the north bank in the east. So far there have been no major jurisdictional disputes between Brandon and Cornwallis but the potential for such certainly exists. An added complication is that water resources are a provincial responsibility, as illustrated by the fact that the city has to consult the province in order to raise the weir at Third Street. Considerable resentment towards the province resulted from its 1987 decision to charge the city for water taken from the Assiniboine, particularly in view of the fact that Winnipeg does not have to pay the province for its water because it comes from an international water body (Shoal Lake) over which Manitoba has no jurisdiction.

Finally, the federal government also has jurisdiction with respect to navigation on rivers and, while the Assiniboine is no longer a navigational route, the case for federal jurisdiction could be argued on the basis that it is navigable. There is then the potential for jurisdictional disputes involving all three levels of government.

**Natural Changes and Human Modifications**

The Assiniboine is a typical meandering river gradually shifting its course over the years. The original survey map, dating from 1881, clearly shows the position of the Snye, which was then one branch of the river channel; and some of the abandoned meanders are also indicated. One of them, located on the golf course, was part of the main course of the river in 1881 (Figure 2:5). Virtually all of the meander loops have become more accentuated since the first map of the area was surveyed.

In addition to these natural changes, barriers have been built across the river at three locations: (1) near the water treatment plant, to divert water towards the intake; (2) a weir at Third Street to lift the level of the water to ensure adequate height at the water treatment intake; and (3) near the hydro plant to raise the water for intake there. At one time, in 1910, there was talk of building a dam across the river at Currie's Landing to generate power at the rapids

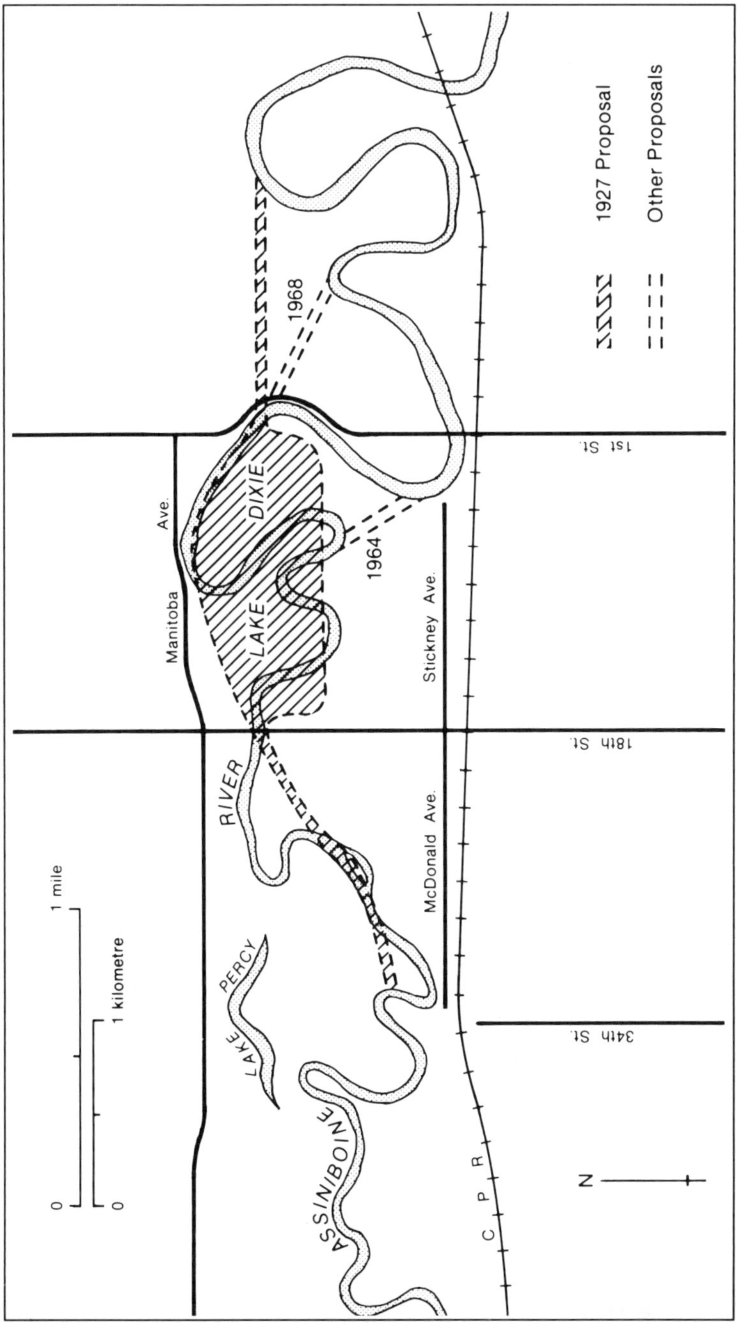


FIGURE 2:6
*Suggested diversions of the Assiniboine River at Brandon*

but nothing came of this, perhaps because the city was already being supplied with electricity from the first hydro plant in Manitoba, located on the Little Saskatchewan River about two kilometres north of its junction with the Assiniboine. Operated by the Brandon Electric Light Company Ltd., the plant served the city via an eleven-thousand-volt wood pole line. The plant, which operated for eight months of the year, was dismantled in 1924 and the dam was eventually swept away in the summer flood of 1949.

There have been several suggestions for rerouting the river but only one of them has been put into effect — the blocking of the Snye. Straightening of meanders as a flood control method is standard practice and has been suggested several times for the Assiniboine at Brandon. One proposal in 1927 called for a straightening of the meanders between Twenty-Sixth Street and Eighteenth Street East with the creation of a lake (Lake Dixie) between Eighteenth Street and First Street (Figure 2:6). The three-mile lake frontage would have had "the most desirable residential sites in Brandon" (*Manitoba Free Press* 15/11/1927). This proposal came to nothing but the idea of shortening and straightening the river was not forgotten and in 1964 a diversion from McGregor Avenue to near the Third Street weir was considered. Yet another alternative east of First Street was considered in 1968 at the time of a long debate over the form and location of the Eighth Street bridge over the CPR lines.

**Acknowledgements**

Most of the information included in the section on flooding was collected by Jeff Dechka for an undergraduate thesis as part of the requirements for a four-year Specialist degree in Geography at Brandon University. Much of the information for the other sections was collected by the late Elizabeth Ann Bremner who was working on an undergraduate thesis on Brandon and the Assiniboine at the time of her death in 1984.

**References**

Barker, G.F. 1977. *Brandon: A City 1881-1961.* Altona: Friesen and Sons.

*The Brandon Daily Sun*, 7 May 1923, 18 June 1927, 16 May 1955.

*Brandon Sun*, 3 August 1988, 5 August 1988, 9 August 1988.

Brown, R. 1981. *Steamboats on the Assiniboine.* Brandon: Tourism Unlimited.

City of Brandon. 1974. *Westbran Work Activity Project, Land Use Study.* Brandon: City of Brandon.

*The Commercial,* 10 October 1882, 24 July 1883.

Dechka, J.A. 1981. Spatial and Temporal Distribution of Flooding at Brandon, Manitoba. B.Sc. thesis, Department of Geography, Brandon University.

Glenn, C.L. 1977. Fishes of the Assiniboine River. *Wildlife Crusader* Summer:10-13.

Johnston, W.A. 1946. *Glacial Lake Agassiz, with Special References to the Mode of Deformation of the Beaches.* Ottawa: Geological Survey of Canada, Bulletin 7.

*Manitoba Free Press,* 15 November 1927.

Mudry, H., G.H. Mackay and V.M. Austford. 1981. Flood control and flow regulation problems on the Assiniboine River. *Canadian Water Resources Journal* 6(4):157-73.

Province of Manitoba. 1976. *Background Studies: Brandon Development Plan.* Brandon: Province of Manitoba, Municipal Planning Branch.

Province of Manitoba. 1985. *Assiniboine River Canoe Route: Journey of the S.S. Alpha.* Winnipeg: Province of Manitoba, Manitoba Surveys and Mapping Branch.

Tyman, J.L. 1972. *By Section, Township and Range: Studies in Prairie Settlement.* Brandon: Assiniboine Historical Society.

Ward, R. 1978. *Floods: A Geographical Perspective.* London: The Macmillan Press Ltd.

# THREE
# The Climate of Brandon and its Surroundings

*Rod McGinn*

In recent years public interest in climatology has grown considerably. Associated with this growth has been an increased interest in urban climatology and agrometeorolgy, the development of user-oriented indices in both these subdisciplines, and the employment of innovative weather broadcasting techniques on television. Urbanites have become aware that large metropolitan centres develop their own local climates. Weather broadcasts which originate from these large cities have shifted their focus towards an audience which inhabits the city core, industrial parks, and the residential suburbs. Terms such as comfort index, humidex, wind chill factor, and pollution index have become an integral component of urban weather reports and forecasts.

In smaller urban centres, which remain closely associated with their agricultural hinterlands, weather broadcasts have assumed a dual focus, serving both the urbanite and the inhabitant of the nearby rural community. Estimates of probability of precipitation, frost warnings and various severe weather warnings usually accompany the daily lists of high and low temperatures. In some agricultural regions, where the city is in fact an agricultural service centre, special farm weather reports are regularly included in the local weather forecast. In these reports additional information on the movement of fronts, squall lines and thunderstorm activity is often provided. Weekly features include soil moisture and crop moisture indices and current observations regarding the progress of the growing season.

Brandon is representative of the small city with a large agricultural hinterland acting as a service centre for numerous rural communities. The climate of the surrounding area has a profound effect on the day-to-day activities within the city and on its long-term fortunes. For these reasons this chapter describes the climatology of Brandon and its surroundings.

Two official climatological data collection sites are located in Brandon. The Canada Department of Agriculture (CDA) station is located at the Agricultural Research Station on the Assiniboine River floodplain, and the Atmospheric Environment Service (AES) weather station at Brandon Airport north of the Assiniboine Valley. Both climatological data collection sites have been in continuous operation since 1951 and they meet the general requirements established by AES for generating climatological normals (the arithmetic mean value of regularly observed meteorological elements, such as temperature, for a specified period of time). In Canada, as in most other countries, the required time period is thirty years, specifically from 1951 to 1980.

## The Seasons at Brandon

The climatic regime of a particular locale is usually subdivided into seasonal components, theoretically each season being based on several biometeorological parameters which influence the growth and development of the natural vegetation. In the continental mid-latitudes four more or less distinct seasons are recognized. Unfortunately, it is popular to assign a three-month period to each season resulting in biometeorological definitions being abandoned in favour of temporal definitions. In this chapter the seasons are defined according to biometeorological parameters.

### *Winter*

In Brandon, winter is the "hibernal season" — a season of plant dormancy. It is the cold season of the year with mean daily temperatures persistently below 0°C. Dominated by anticyclonic activity, it is relatively dry with precipitation being almost entirely in the form of snow. Given this operational definition, winter in Brandon is a five-month period from the beginning of November until 5 April (Figure 3:1).

During this relatively long winter season Brandon receives an average of three hours and fifty minutes of bright sunshine each day. The associated input of radiant energy (only 0.243 megajoules per square metre per day) is converted into a small quantity of sensible heat which is reflected in the low ambient temperatures recorded at the Brandon climatological stations. Minimum daily temperatures are often well below -15°C and mean monthly

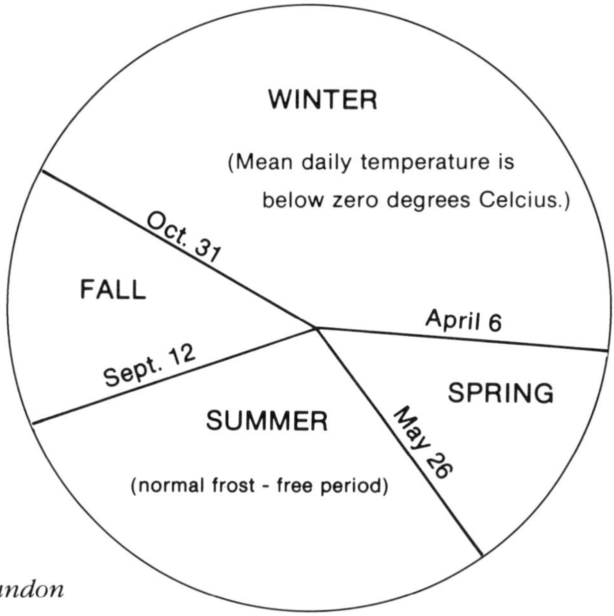


FIGURE 3:1
*The seasons for Brandon*

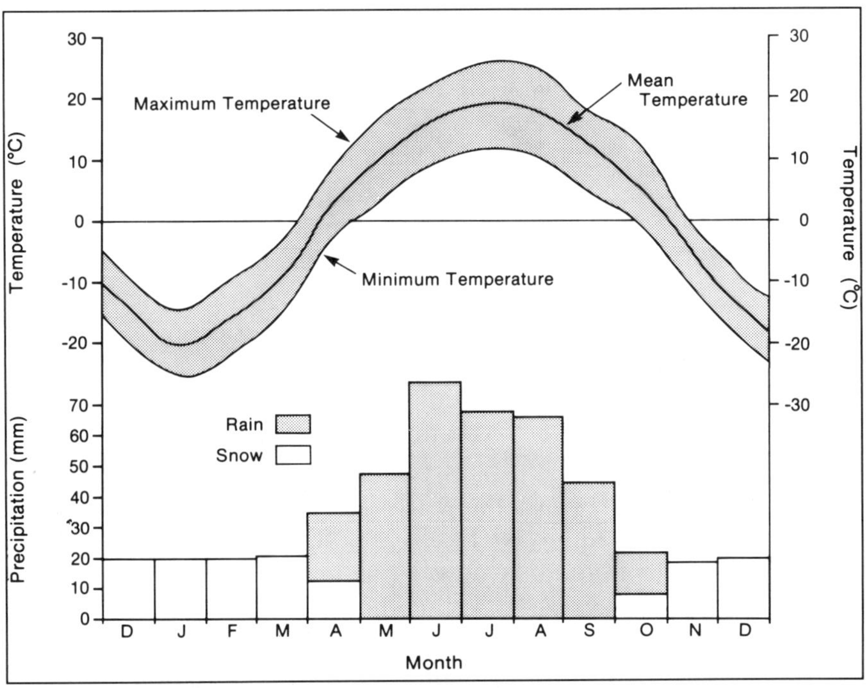


FIGURE 3:2
*Climatic diagram for Brandon*

maximum temperatures have never been above 0°C (Figure 3:2) although daily values can exceed the freezing point, as they frequently did during the warm spell in late December 1986 and early January 1987.

In 1966, Brandon experienced the coldest January in the ninety-six years that records have been kept, with the average daily temperature recorded at the Agricultural Research Station being -27.5°C and the minimum daily temperatures on 7 January and 28 January being measured at -45.5°C. This temperature is only 0.6°C warmer than the all-time low of -46.1°C recorded on 13 January 1916. In the same month, fifty kilometre winds dropped apparent temperatures to -68°C, causing several power failures when hydroelectric insulators shattered in the extreme cold.

Relative humidities during this season are deceptively high — 77 percent on average throughout the winter. However, the cold air contains only small quantities of water vapour, so both the specific humidity and the vapour pressure are very low, averaging only 1.45 grams per kilogram and 0.23 kilopascals respectively.

Snowfall is evenly distributed throughout the five-month season, averaging near twenty millimetres (water equivalent) each month (Figure 3:2). Normally most of the monthly snowfall takes place during one to three prominent snowfall events which are always associated with synoptic scale cyclonic storms, each of which persists over the region for three to five days.

At both the beginning and the end of the winter, some precipitation falls as freezing rain. On average Brandon receives eight days of recordable freezing rain each winter. Normally these precipitation events are relatively insignificant, occurring once or perhaps twice in any given winter month. However, approximately every second year a more severe freezing rain storm affects the region; for example, on 5-6 March 1983, twelve to twenty-five millimetres of ice (in the form of freezing rain) was deposited on the streets of Brandon. The weight of this ice, and that of a centimetre of snow accompanied by twenty-five to thirty kilometre per hour winds, resulted in numerous power outages as hydroelectric transmission lines and poles collapsed. Radio and television communications were also disrupted when three broadcast towers collapsed, the most prominent tower being the four-hundred-metre, one-million-dollar Hayfield transmission tower located on the western edge of the Brandon Hills. The two Brandon television stations and four FM radio stations were put off the air for several days.

Winter winds are usually from the west or west-northwest and average approximately sixteen kilometres per hour. However, the maximum monthly wind velocities are very high (seventy-eight kilometres per hour) and wind gusts can exceed one hundred kilometres per hour. Only the springtime winds are stronger. Surprisingly the winter months, in particular January and February, also experience high monthly percentages of calm. Calm atmospheric conditions often occur at sunrise when very cold Arctic air overlies the Brandon region. Such weather conditions favour the development of two unusual optical phenomena known as sun dogs and sun pillars.

Sun dogs, or parhelia, are two reddish luminous spots that can be seen on both sides of the sun. They are usually located at an angle of twenty-two degrees from the sun and at the same elevation. Parhelia are produced by the *refraction* of sunlight in hexagonal ice crystals which are falling with their principal axes oriented vertically, a condition that can only occur in relatively calm cold air. When the wind velocity increases, the ice crystals change their orientation and the sun dogs disappear.

A sun pillar is often observed at the same time as sun dogs. This luminous streak of white or reddish light extends directly above and below the sun. In this case, the luminosity is believed to be produced by the *reflection* of sunlight from the tops and bottoms of the columnar hexagonal ice crystals which are falling with their principal axes vertical. Once again, calm cold air is essential.

The blizzard is probably the most dramatic and hazardous meteorological phenomenon associated with the winter season on the Canadian Prairies. Blizzards are defined as severe weather conditions, characterized by low temperatures and strong winds (at least fifty kilometres per hour) which carry sufficient quantities of snow in the air to reduce visibility to less than 150 metres. Severe blizzards have winds which exceed seventy kilometres per hour, temperatures below -10°C, and visibilities which are near zero. No statistics exist on the occurrence and characteristics of blizzards in the Brandon region, but there is some information about visibility and the occurrence of blowing snow.

In southwestern Manitoba blowing snow conditions are responsible for numerous travel advisories and the occasional closure of highways. Coupled with low temperatures and high wind chill effects, blowing snow can produce hazardous and potentially life-threatening situations. Brandon Airport records a seasonal average of seventeen days when visibility is significantly reduced by blowing snow, on average four of these days occurring in December and another six in January. The nearby rural areas experience even more days of blowing snow; for example, the climatological station at Rivers Airport, located thirty kilometres northwest of Brandon, records twenty-eight days of blowing snow each winter.

The very low temperatures, the wind chill, and the overall length of the season combine to give Brandon a winter discomfort factor of 41 out of a possible 70 (Phillips and Crowe 1984). While this relatively high value is slightly greater than the mean for the prairie provinces, it is still one of the lowest in Manitoba, suggesting that Brandon is one of the better places in Manitoba to live during the winter.

*Spring*

Spring, or the vernal season, is the time of the year when the transition from winter to summer takes place. Mean temperatures rise above 0°C and the cyclonic activity common in the late winter diminishes. Meteorologists usually consider spring to include the months of March, April and May, but, as March has already been identified as a winter month, spring in the Brandon area is a two-month transition period from 6 April to 25 May (Figure 3:1).

As the number of daylight hours increases, so does the number of hours of direct sunlight, April and May averaging 7.5 hours of direct sunlight on each day (Figure 3:3). A large proportion of the solar radiation during this season is used to melt the annual snowpack, to warm the ground and to heat the air. The average -5°C overnight lows of early April soon rise above 0°C and reach 6°C by late May. Similarly daily maximum temperatures rise from 3°C in April to 20°C by the end of May (Figure 3:2). Despite this surge in daily temperatures, frosts usually occur in both April and May.

Over 80 percent of the total springtime precipitation (eighty-four millimetres) falls as rain. Nevertheless, significant amounts of snow can fall during this season. It was during a spring snowstorm that the maximum twenty-four hour snowfall occurred — on 25 May 1961

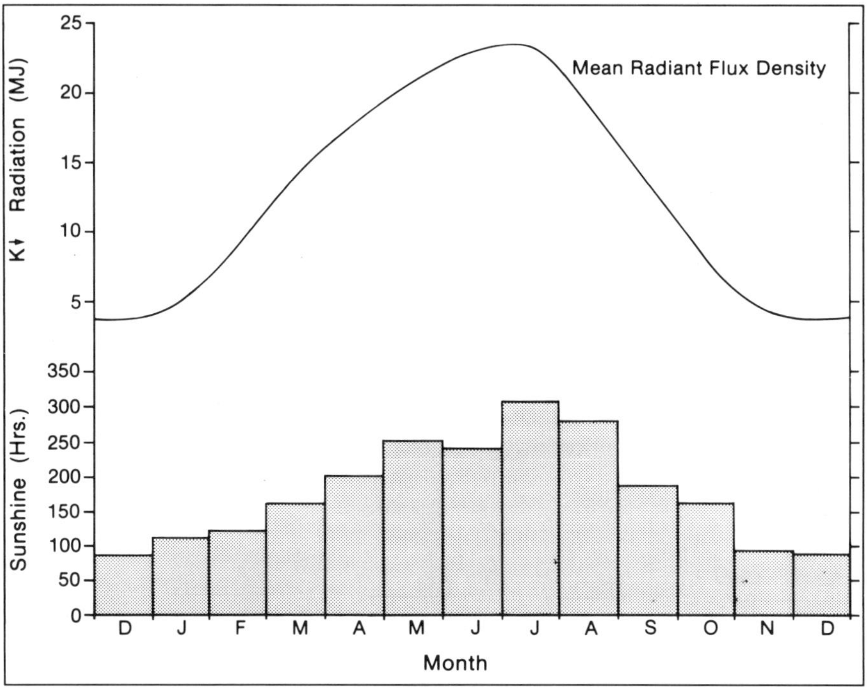


FIGURE 3:3
*Radiation and sunshine hours for Brandon*

an incredible forty-seven centimetres fell during a twenty-four-hour period.

Once the surface of the ground thaws, most of the rain and melting snow is used to recharge the depleted soil moisture to values near field capacity — assumed to be two hundred millimetres in the Brandon region. During the 1980 drought the spring rains did not come. In fact for almost sixty days (27 March to 2 May) there was no precipitation recorded at both Brandon climatological stations and by mid-June the region was acknowledged to be "one of the three most severe drought stricken areas of the world" (Environment Canada 1980).

The spring winds are probably the most significant component of the annual wind regime for Brandon. During April and May the prevailing wind direction is from the east-northeast and when the easterly and northeasterly winds are included, on an average they blow for 29 percent of the time. North-northwest, northwest and west-northwest winds are also common although the frequency of occurrence is considerably less than the east-northeast winds. Spring winds have the strongest mean monthly velocities (approximately eighteen kilometres per hour) and winds from the prevailing directions always average more than twenty kilometres per hour. The average maximum wind velocities are greater than eighty kilometres per hour with gusts that exceed ninety kilometres per hour. At this time of year, many of the agricultural fields lie more or less exposed to the wind, resulting in frequent dust storms and extensive soil erosion. Agricultural activities associated with seedbed preparation add to the problem.

*Summer*

Summer in southwestern Manitoba is truly the "aestival season." Many small towns take advantage of the long hours of daylight, bright sunshine, high temperatures and low humidities to stage their summer fair.

Summer is also the normal frost-free period; that is, the 104-108 days when the probability of frost is less than 50 percent, specifically from 26 May to 12 September. However, the length of the frost-free season varies greatly from year to year, the longest frost-free period ever experienced lasting 144 days from 7 May to 19 September while the shortest period on record was only 55 days, from 3 July until 28 August. First frosts have occurred as early as 3 August and as late as 9 October and last frosts have been as late as 3 July and as early as 30 April. Clearly, frost can occur, and has occurred, in every month of the normal growing season — an ever present climatic hazard which has a significant effect on agricultural activities in and around Brandon; for example a severe frost on 21 May 1987, which came after a long period of warm weather resulting in early seeding, caused considerable damage to crops in the region.

In the Brandon area, summer includes the traditional three-month period of June, July and August. During this period, days are relatively long, averaging 16.5 hours. An average of 823.7 hours of bright sunshine is recorded during the three-month period, a value equivalent to nine hours of direct sunlight each day (Figure 3:3). The associated shortwave energy input of 0.704 megajoules per square metre per day is almost three times greater than the comparable winter statistic. This high amount of radiant energy is responsible for the growth and rapid maturation of the local agricultural crops and garden vegetables.

Summer daily maximum temperatures are approximately 25°C on average while overnight lows average 10°C (Figure 3:2). Extreme maximum temperatures have exceeded 37°C within the last thirty years, while at the other extreme at least one frost has been recorded in each of the summer months in the period since 1951.

Although relative humidities are generally low in the summer months, specific humidities and the associated vapour pressures are relatively high (8.58 grams per kilogram and 1.36 kilopascals respectively). Partly as a consequence of this high water content, the mean atmospheric pressure (adjusted to sea level) is a relatively low 101.31 kilopascals compared to the winter average value of 102.0 kilopascals. Although the difference between these two pressure values seems insignificant, it is important to aviators. Atmospheric pressure is a function of the mass density of air and the more dense the air the more efficient it is in supporting an aircraft. Since the moist summer air is less dense than the drier winter air, a pilot will require, on average, a significantly longer runway for take-off during the summer months than during the winter.

Approximately 210-220 millimetres of precipitation fall on the Brandon area during the summer. This value represents 45 percent of the annual total and three and a half times the winter snowfall accumulation. Although measurable rainfall is recorded on an average on ten days each month, the June total precipitation (seventy-nine millimetres) is significantly greater than the July and August totals (sixty-eight and sixty-seven millimetres respectively). A large proportion of the summer rainfall is associated with convectional activity reflected in the fact that Brandon averages seventeen thunderstorms each summer and at Rivers Airport an average of twenty-two thunderstorms is recorded during the summer season.

June rainfall provides the necessary moisture for seed germination and early plant growth whereas periodic July and August rains sustain growth and reduce plant stress caused by moisture deficiency. This is not to say that the Brandon region does not experience dry spells and drought conditions. On the contrary, the temporal and spatial variability of convectional rainfall ensures uncertainty in the summer rainfall regime and moisture conditions can range from excessively wet to severe drought. In recent years severe drought conditions occurred in 1961, 1967 and 1980, whereas 1971, 1975 and 1982 are representative of moderately wet summers.

The extreme drought conditions which persisted throughout the spring and summer of 1980 were largely the result of several agroclimatic conditions which have led to crop failures in the past. The soil water reserves were very low in March 1980 (75 percent of normal) because they had not been recharged by autumn rainfall in 1979. The 1979-80 winter season was itself drier than normal, consequently there was little snowmelt available to assist in soil water recharge. The climate of the spring season was dominated by an intense, persistent high pressure cell which brought very warm, dry, stable air into southwestern Manitoba. Maximum daily temperatures were 15°-18°C above normal, consistently over 25°C and occasionally over 30°C. There was no rain in April and only eight millimetres of precipitation in May. This general climatic trend continued and by mid-May it was recognized that the top soil zones did not contain sufficient moisture to initiate hay seed germination. From 70-90 percent of the expected hay tonnage was lost. Cereal crops are planted deeper in the soil and benefitted from drought-abating rains in late June and early July. However, it is estimated that 48 percent of these crops also failed.

In retrospect the major reasons for drought in the Brandon region appear to be associated with unusually low soil water reserves, warm dry spring seasons and the development of a persistent high pressure cell over the area. The combination of the "Omega Block" (the high pressure cell) and reduced plant-available water (precipitation and soil water) causes the moisture deficit period (see the section of this chapter on the Climatological Water Balance) to be set back from the late summer to late spring and early summer. This results in water-deficit-induced stress during the early stages of crop development. Since the effect of water-deficit stress on grain crops is acknowledged to be cumulative in nature, the earlier the crops are subjected to the stress the greater the potential for damage. In 1980 it appears that the agricultural crops of the Brandon region experienced this stress one full month before normal.

The summer winds are variable in their direction; westerly and west-northwest winds are the most common, but easterly and east-northeast winds are almost as frequent. The average wind velocity during the summer is a relatively low 14.9 kilometres per hour.

However, the average maximum monthly wind velocities are greater than seventy kilometres per hour. Strong winds may come from several directions; for example, west-northwest, north-northwest, and south-southeast. This variability is not surprising, because sudden and violent wind shifts are commonly associated with the unstable atmospheric conditions which exist periodically during the summer. Unstable atmospheric conditions often result in thunderstorms during which wind gusts have exceeded one hundred kilometres per hour and hail is a possibility. At such times tornadoes can occur, but rarely do. The high intensity rainfall and lightning associated with the thunderstorms also pose a threat to human activities.

Periodically, during the summer season, intense pressure gradients develop in western Canada. This synoptic situation can generate widespread and prolonged winds of high velocity. In the past, these winds have caused considerable damage to both agricultural land and urban property. On 10 June 1985 a severe regional windstorm occurred over southern Saskatchewan and southwestern Manitoba. Average wind velocities were estimated to have been between seventy and one hundred kilometres per hour with gusts recorded at 130 kilometres per hour. These near-hurricane force winds uprooted trees, stripped shingles from rooftops and brought down the southeast wall of Brandon's Commonwealth Air Training Plan Museum. In July 1970 a similar storm produced the highest gust velocities ever recorded in Brandon (137 kilometres per hour). Fortunately, such severe windstorms occur only about once every fifteen years.

Phillips and Crowe (1984) estimate the summer discomfort index for the Brandon region at 10 out of a possible 30 points. This estimate is the lowest for any climatological station in Manitoba; it is slightly above the average for western Canada, but below the indices for all the major urban centres in eastern and central Canada. In short, from a climatic perspective, Brandon is a pleasant place to live during the summer.

### *Autumn*

Like spring, autumn is a transitional season and like spring the transition is a relatively short period; from a climatological perspective, autumn begins on 12 September and concludes on 31 October (Figure 3:1). The first frost normally occurs on

12 September and the overnight lows are usually less than -5°C by the end of October. Also, mean daily temperatures decrease during the autumn reaching 0°C by 31 October. However, the average maximum daily temperatures are still slightly above 0°C by the end of the season. September's forty-seven millimetres of precipitation is almost entirely rainfall and approximately 75 percent of the October precipitation total (22.45 millimetres) also falls as rain, but on average Brandon experiences at least one snowfall in October. For example, on Thanksgiving weekend in October 1959 Brandon and all of southwestern Manitoba experienced a severe snowstorm of near blizzard proportions. Eighty-four centimetres of snow fell in a three-day period (8-10 October) and fifty kilometres per hour winds caused heavy drifting, at times reducing visibility to zero. That year the snow did not melt immediately and the ground was snow-covered from October until the next spring.

The fall winds are predominantly from the west and west-northwest. The mean wind velocity is 16.4 kilometres per hour with an average maximum wind velocity of approximately seventy kilometres per hour.

## Climatological Elements

### *Solar Energy and Sunshine*

Data on hourly shortwave radiant energy on a horizontal surface recorded at Winnipeg International Airport are taken as representative of the shortwave energy regime at Brandon, which is at approximately the same latitude and only two hundred kilometres to the west. Mean values are lowest in the winter with a minimum of 3.82 megajoules in December, and highest in the summer with a maximum value of 22.99 megajoules being recorded in July.

The potential and actual hours of direct sunshine for each month are recorded at Brandon CDA climatological station. The ratio of actual to potential hours of direct sunshine, expressed as a percentage and termed the "sunshine ratio," is a useful climatological statistic worth detailed examination. Brandon has an annual average sunshine ratio of 44.8 which means that it receives approximately 45 percent of its shortwave radiant energy as direct sunshine, a relatively high value compared to other Canadian centres. The lowest sunshine ratios are in the winter months, the seasonal mean of 34.4 percent being well below the annual average. It is interesting to note that the ratios for both November and December (30.6 percent and 28.0 percent respectively) are very low, while the January, February and March values are 6 to 8 percent higher. This is probably due to the greater frequency of cyclonic storms during the early winter period and the general dominance of a ridge of Arctic air during January, February and early March.

The monthly sunshine ratios rise through the brief spring transition period giving a seasonal mean value of 50.3 percent. The 460.6 hours of direct sunshine received during this two-month spring are only 120 hours less than the total received during the preceding five-month winter season. More importantly, the total springtime radiant energy (38.64 megajoules per square metre) recorded in two months is greater than the total radiant energy received during the five-month winter (36.82 megajoules per square metre). This means that on average each square metre of land receives two and a half times more radiant energy for each spring month resulting in the thawing and drying of the agricultural land around Brandon. Also, the significant increase in daylight hours and in the hours of bright sunshine, fifteen hours and 7.5 hours respectively, undoubtedly has a positive effect on the psyche of both the urban and rural populace.

The sunshine ratio reaches a maximum of 62.2 percent in July, almost 8.5 percent above the seasonal average and 17 percent greater than the mean annual value, representing a mean total of 308.4 hours of bright sunshine for July. August contributes an additional 273.3 hours and also exhibits a sunshine ratio greater than the summer mean of 53.8 percent. However, the June sunshine ratio is considerably lower than might be expected, the monthly mean value of 44.8 percent being less than the springtime average and only 1 percent greater than the autumn mean. At Winnipeg International Airport the ratio is 5 percent greater than the Brandon ratio and is closer to the expected value for the summer season. The unusually low June ratio recorded at Brandon CDA is probably a consequence of the high frequency of convective activity in the Brandon region during the month, reflected in the fact that the Brandon Airport climatological station reports on average five days of thunder for the month of June, one of the highest values recorded in Canada for a given month. Convective activity not resulting in thunder is even more common and the associated cloud forms undoubtedly prevent direct sunlight for long periods in the late afternoons and early evenings. Although an average of only 242.0 hours of bright sunshine is recorded for the month of June at the Brandon CDA climatological station, the summer mean total of 823.7 hours represents 40 percent of the annual average.

As cyclonic activity increases during the autumn, the hours of direct sunlight decrease to an average of 184.1 hours in September and 160.2 hours in October. The sunshine ratio also decreases to a seasonal average of 43.5 percent.

*Temperature*

The average daily maximum, minimum and mean temperatures at a specific site depend on the mean radiant energy incident on a square metre of land at that site. Therefore, it is not surprising to find that the maximum, minimum and mean temperature regimes for Brandon are highly correlated with the solar energy regime estimated for the area. As this is the case, the temperature regimes have been modelled as a sine function having an annual period (Figure 3:2). In Brandon the highest values for both the maximum and minimum temperature normals are recorded in July and the lowest values occur in January. The highest normal maximum temperature is approximately 26°C; the lowest normal maximum temperature is -15°C giving a normal range of maximum temperatures of 41°C. The extreme maximum temperatures ever recorded at Brandon were 37.8°C at Brandon Airport, and 43.3°C at the Brandon CDA climatological station. The highest average minimum temperature occurs in July (approximately 13°C), the lowest being in January (-26°C), giving a range of 39°C. The extreme minimum temperatures ever recorded in Brandon are -45.6°C at Brandon Airport and -46.7°C at Brandon CDA.

The temperature regime illustrates the well-known temporal lag between receipt of solar radiation and the thermal response to that radiant energy. When dates of maximum temperature and minimum temperature are compared with the respective dates of maximum and minimum radiant energy, a twelve-day lag between shortwave radiant energy and temperature response is observed. The temperature plot (Figure 3:2) was used to separate the extended winter season, defined previously as the hibernal season with mean daily temperatures persistently below 0°C, from the spring and autumn transitional seasons. Normal maximum daily temperatures do not rise above 0°C until approximately 27 March and minimum daily values do not cross the 0°C line until late April. As mean temperature is mathematically defined as the maximum plus the minimum divided by two, the normal mean daily temperatures are persistently below 0°C throughout March and into April, rising above 0°C on or about 6 April. Thus March is clearly a winter month in the Brandon area (Figure 3:1). A similar analysis indicates that the normal maximum temperature drops below the 0°C line in mid-November, that minimum daily temperatures are below 0°C by early October, and that the mean daily temperature normally is below 0°C on 31 October. Thus November is included in the winter season.

This type of graphical and statistical analysis of the normal temperature regime for a specific site can be useful in advancing generalizations about the climate of the site. However, it can also lead to disastrous oversimplifications as illustrated by an attempt to use the method to establish the expected frost-free period for Brandon. Frost occurs when the overnight low, or minimum temperature, drops below 0°C. Thus it could be argued that the expected date for a first frost would be when the normal minimum temperature drops below 0°C; that is, on 12 October. Similarly, the date of the last frost should occur when the normal minimum temperature rises above 0°C (27 April). If this were in fact the case, Brandon would experience an average frost-free period of 168 days, not the documented 104 and 108 days recorded at the climatological stations.

*Degree-days*

A "degree-day" is defined by Atmospheric Environment Services Canada as a measure of the departure of a temperature (usually the daily mean temperature) for one day from a given standard or reference temperature. One degree-day is equivalent to each degree Celsius of departure of the mean daily temperature from the standard reference temperature. Various reference temperatures are used. In this chapter, 18°C will be employed when discussing "heating degree-days" and 5°C will be the reference temperature for the discussion of "growing degree-days."

Degree-days are usually accumulated through a specified time period or season. As this is the case, at any time during the "season" the accumulated total can be used as an index to relate temperature to some other measurable quantity such as fuel consumption, energy requirements or plant growth. For Brandon Airport the annual total of heating degree-days is 6,115.8, 97.5 percent of these occurring in the fall, winter and spring seasons — 76.3 percent in the winter alone. Thus fuel consumption indices usually begin in the fall period and carry through until the end of May.

The growing degree-day data for Brandon Airport clearly indicate that the actual growing season cannot possibly begin in March and is very unlikely to begin even in April. The total number of growing degree-days for the Brandon region is approximately 1,650 in the city; along the Assiniboine River the number is slightly higher. Seventy percent of the growing degree-days occur during the summer, 10 percent in late spring (May), and an additional 10 percent in early September. Growing degree-day statistics are of considerable value to farmers, since the requirements for the growth stages of most field crop species can be given as a number of growing degree-days. Thus a farmer can use a degree-day index or total for developing a planting schedule, predicting the various stages of plant growth, and for forecasting the harvest times for each of his crops. Similar timetabling can be used by rural and urban horticulturalists.

*Precipitation*

The winter precipitation (approximately one hundred millimetres) is evenly distributed throughout the five winter months, averaging approximately twenty millimetres of water equivalent per month, virtually all of it falling as snow (Figure 3:2). Each summer month averages at least three times the monthly precipitation of each winter month, all of the precipitation falling as rain or hail. June is the wettest month, receiving on average approximately seventy-seven millimetres of rain, the maximum monthly rainfall ever recorded at Brandon falling in June (141.0 millimetres at Brandon CDA).

The mean number of days of measurable precipitation recorded each month at the Brandon CDA station reveals that a relatively even distribution of precipitation events occurs throughout the year but this statistic does not reflect the nature of the precipitation events with respect to duration, intensity and total precipitation received — all of which are variable.

The winter season averages seven days of snowfall each month. These snowfall days are usually associated with large-scale cyclonic storms which may influence the local area for several days. Thus one snowfall event may be responsible for anything from one to five days of snow. Consequently, for the winter season, the average five-day precipitation total may be a more meaningful statistic. Throughout the winter months, Brandon averages 3.5 millimetres, water equivalent, of snow every five days; the comparable statistic for rainfall during the summer season is four times greater (14.3 millimetres of rain per five days).

Summer precipitation is not always associated with large scale cyclonic storms; often local, and more intense, convectional storms account for a day of rain. Furthermore, the inherent variability of these storms with respect to duration, rainfall intensity, areal extent and precipitation total preclude the use, as meaningful descriptive tools, of statistics such as the five-day precipitation normal and the number of days of precipitation. Keeping this in mind, it can be stated that each summer month averages ten days of rain — a statistic which, to some extent, indicates the frequency of convectional activity in the Brandon region. However, it does not represent the number of convectional storms as several storm cells can, and often do, occur on the same day and late night storms are recorded on two separate days. Suffice it to state that summertime rainfalls are commonly associated with both cyclonic and convectional storms while winter snowfalls are almost always generated by synoptic (large) scale cyclonic storms. Rainfall resulting from convective activity probably accounts for the greater variability in the summer monthly precipitation totals. Standard deviations about the mean range from thirty-seven to fifty millimetres, while the coefficient of variability equals 1.54. The variability of winter snowfall is slightly less; standard deviations average 13.5 millimetres, and the coefficient of variability is 1.41.

Precipitation during the spring and autumn months is significantly greater than that received during the winter months but less than is received during each of the summer months. Most of May's forty-eight millimetres of precipitation falls as rain while the April precipitation total of thirty-five millimetres normally includes twice as much rain as snow; that is, the snow/rain ratio equals 0.53. Similarly, September's forty-seven millimetres of precipitation is almost entirely rainfall, whereas the October snow/rain precipitation ratio averages 0.35 for the two Brandon climatological stations.

*Winds*

The annual wind regime for Brandon and its surrounding areas is difficult to assess except in general terms. This is due to the turbulent nature of air flow, the associated variability in wind velocities, and a high degree of variability in wind direction. Other difficulties arise from the relatively short period of records and technical problems associated with the measurement of wind velocity.

Average monthly wind velocities for the sixteen wind vectors can range from a low of approximately 10.7 kilometres per hour (from the south in February) to a high of 23.3 kilometres per hour (from the northwest in January). The mean wind direction is with two exceptions from the west, these exceptions occurring in the spring, when the prevailing wind direction for both April and May is from the east-northeast. These east-northeast winds also have the strongest mean monthly wind velocities experienced, averaging 18.1 kilometres per hour.

Despite being the strongest *mean* monthly winds recorded, east-northeast winds seldom represent the average *maximum* wind velocities measured in any given month. In fact, 50 percent of the time the average maximum monthly wind velocity is associated with a west-northwest wind. This figure rises to over 80 percent when the other westerly wind vectors are included. Mean monthly maximum wind velocities range from a recorded low of sixty-nine kilometres per hour in September to a high of eighty-four kilometres per hour in February. In other words, on average Brandon can expect to experience maximum wind velocities each month which will exceed sixty-nine kilometres per hour. Generally the mean maximum wind velocities recorded during the summer and autumn months fall into the lower part of the range, below seventy-six kilometres per hour, while the winter and spring winds commonly exceed the average and lie in the upper part of the range.

The highest wind velocities ever recorded (gust velocities) are also mainly associated with westerly wind vectors. Gust velocities of over one hundred kilometres per hour have been recorded during several months — January, February, June, July, October and November. The extremely strong gusts which occur in the summer are associated with intense convective activity (thunderstorms), the maximum gust ever recorded, one July, being estimated at 137 kilometres per hour.

As already indicated, west is the most common wind vector in every month except April and May; west-northwest winds are usually the second most frequent winds. In fact, during the autumn and winter months, west and west-northwest winds occur at least 25 percent of the time and when northwest winds are included the value rises to a monthly average of 38.7 percent. During the spring, east-northeast winds predominate and when easterly and north-easterly winds are included these winds occur an average of 29.0 percent of the time throughout April and May. Summer winds are quite variable in direction, west-northwest winds usually being the most common with east-northeast winds almost as frequent, this variability being associated with the unstable atmospheric conditions which exist during the thunderstorm-prone summer season.

Days of calm are most frequent in January and February when anticyclonic activity is at its maximum. During these days, when very cold, dry Arctic air lies over Brandon, it is said to be "so cold that even the wind doesn't blow." High pressure cells can also prevail during July and August, undoubtedly contributing to the relatively high percentage of calms during the late summer.

## The Climatological Water Balance

The climatological water balance is a climate-based accounting system which evaluates the moisture gains and losses at a particular site. For the Brandon region, accounting was based on the 1959 to 1984 mean daily temperature and precipitation data, following the procedure outlined by Johnstone and Louie (1983). The resulting graph (Figure 3:4) is a plot of fifty-two weekly values.

The first significant snow accumulation likely to remain on the ground occurs in the first week of November (week forty-five) when mean daily temperatures dip below 0°C and winter snow accumulation begins. During the twenty-one-week winter period, snow accumulation averages 4.24 millimetres (water equivalent) each week for a total snowpack accumulation of approximately eighty-nine millimetres.

Snow melt begins in early March, but most of the meltwater probably refreezes within the snowpack and is not readily available for soil moisture recharge or for surface runoff. Significant snow melt usually begins during the last week in March and lasts until 30 April (week eighteen). Most of this meltwater runs off as the spring freshet or infiltrates into the thawing surface layer of the soil.

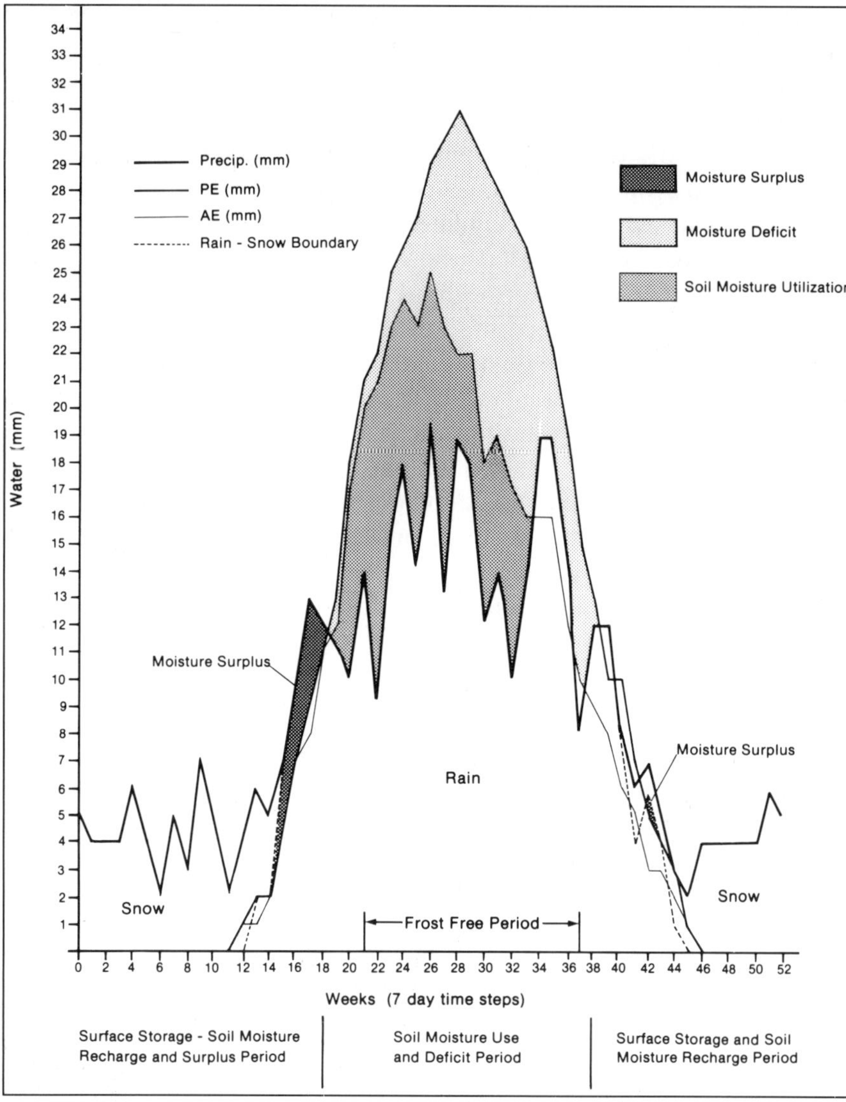


FIGURE 3:4
*The climatological water balance for Brandon*

Springtime rainfalls generally infiltrate into the ground, recharging the soil moisture levels to near field capacity (assumed to be two hundred millimetres in the Brandon area). Of this total, sixty millimetres are held in the upper soil zones and 120 millimetres occupy the pores and capillaries of the lower part of the solum.

Once the unfrozen soil is at, or near, field capacity any excess spring rainfall is designated as surplus moisture and is assumed to run off.

The growing season unofficially begins when the mean daily temperatures are persistently above 0°C, normally during week fourteen (2-8 April) and ends on 31 October, when the mean daily temperature drops below 0°C. Included in the growing season, but not to be confused with this season, is the frost-free period (normally 26 May to 12 September, a 104-day period). During the early part of the growing season, and before the beginning of the frost-free period, farmers and horticulturalists till the fields, prepare seed beds, and plant their crops. Rising temperatures and spring rainfall facilitate seed germination and early plant development.

The weekly water budgets, which explain the plant-moisture requirements and moisture availability, comprise the climatological water balance and are most critical during the frost-free period. At this time, moisture inputs and available soil moisture determine plant growth, maturation rates and, ultimately, crop yields. Any serious, prolonged moisture deficiency, or surplus, can cause plant stress which impedes the normal growth and maturation process. On the Canadian Prairies moisture deficiency is an almost annual occurrence.

Potential evapotranspiration (PE) is the term used to describe the amount of water that would normally be evaporated and transpired from a vegetated surface if there were an unlimited supply of water available to the plants at all times. As a result, PE represents a maximum which depends on several bioclimatological parameters such as temperature, day length and vegetation type, the most important variables appearing to be temperature and day length. Consequently, these two variables are generally used to compute the PE for a station. Fifty-two computed PE values are plotted on Figure 3:4. Maximum water requirements normally occur during week twenty-eight (9-15 July) when PE equals thirty-one millimetres, and minimum values occur during the winter.

The weekly rainfall during the sixteen-week, frost-free period displays considerable variability, ranging from a minimum of eight millimetres to a maximum of nineteen millimetres. Regardless, the normal weekly amount of water input is always less than the respective weekly water requirement (PE). These weekly shortfalls can be satisfied, more or less, by water drawn from the soil.

When the soil moisture storage value is near field capacity it is assumed that the soil moisture withdrawal rate equals the PE rate with the result that the actual evapotranspiration (AE) is almost equal to the PE. Once the soil moisture storage values have been reduced to 60 percent of the field capacity, soil moisture withdrawal rates are reduced and the AE becomes significantly less than the PE. The difference, PE - AE, is termed the moisture deficiency or deficit, and the difference between the AE and precipitation for each time period (AE - Prec.) represents the amount of water withdrawn from the soil by the plants for evaporation and transpiration.

The average weekly AE values for the twenty-five year period, 1959-84, are plotted on Figure 3:4, so the composite graph illustrates the average weekly relationships between precipitation, PE and AE throughout the year. It also indicates the normal magnitude and temporal extent of the various soil moisture conditions in the Brandon area.

A moisture surplus usually occurs in April when the weekly accumulated rainfall totals exceed the agrometeorological water demand (PE) and the fields are saturated. The excess water generally runs off via a network of drainage ditches. However, in some locations where drainage is poor, the annual water surplus can cause tillage problems.

By mid-May, the weekly PE totals exceed the accumulated rainfall and a moisture deficiency arises. During seed germination and the early stages of crop development, moisture held in the upper layer of the soil is withdrawn to meet the rising evapotranspiration demand. This process is known as soil moisture utilization and is considered vital in prairie agriculture. In recent years, when the springtime soil moisture levels have been low, agricultural drought and crop failures have been common. Irrigation practices, when and where they are available, play an important role in alleviating these droughts.

By July and August, most of the readily available soil moisture has been withdrawn and a true moisture deficit condition exists. Fortunately, evapotranspiration rates are declining and the periodic summer rainfalls can usually provide sufficient moisture to sustain crop growth throughout the maturation and heading stages. Again irrigation can assure crop success during particularly dry years.

By mid-September, weekly precipitation totals are once again greater than the agrometeorological moisture demand and a period of soil moisture recharge begins. This is a crucial period for prairie agriculture because both the upper and lower soil layers can be recharged with the moisture necessary for the coming crop year. However, once the ground becomes frozen (week forty-five) the accumulated precipitation cannot recharge the soil and the precipitation (snow) is temporarily stored until the spring melt.

## Conclusion

The annual climatological water balance can be viewed as a measure of the agricultural potential and/or limitations of a place. Likewise, climate-based indices such as the humidex and the wind chill factor can be considered to represent environmental stresses imposed on a region's inhabitants and that region's hazardousness for its people.

Phillips and Crowe (1984) use a climate severity index which summarizes the climate stressors of a place and weighs these factors so that they reflect properly the climate's severity. Four general factors are considered: climate imposed discomfort; hazardousness; outdoor mobility; and psychological state.

The discomfort factor represents a weighted total of comfort sensations for the winter (70) and the summer (30) seasons. For Brandon, the winter season receives a relatively high discomfort value (41) due to the length of the season (157 days), the persistence of wind chill and the extreme cold during January. To some extent, summer comfort sensations offset the winter severity giving Brandon a total discomfort value for the year of 51 out of 100 — still a relatively high value when compared to values for other Canadian cities.

The severity of the climate in the Brandon area makes the city a potentially dangerous place to live. Based on a hazardousness index of 100, Brandon scores a 36, which is higher than 68 percent of the cities in Canada. Winter hazards such as blizzards, snowstorms, blowing and drifting snow, and wind chill play a significant role in determining the value. High winds and summer thunderstorms are also contributors to the hazardousness factor.

Climatic hazards do not influence the outdoor mobility factor, since hazardous events are usually of short duration. Limitations to outdoor mobility in the Brandon area are well below the average (39) for major Canadian centres. Brandon scores 31 out of 100 in this category which is determined by days of snowfall, freezing rain, fog and less than ideal visibility.

The psychological state factor is directly related to the hours of sunshine, number of cloudy days and precipitation events; in other words, the ratio between gloomy days and bright days. From a psychological perspective Brandon is a great place to live, there being only a few major urban centres which can provide a comparable psychological climate.

Once the climatic stressors are properly weighted and totalled Brandon scores a Climate Severity Index value of 49 out of 100 points. For Brandon dwellers, this is a discouraging value, since it is indicative of a relatively severe climate and is one of the highest index values calculated for Canadian cities. However, they can draw some solace from the fact that St. John's, Newfoundland, Quebec City and Winnipeg are all considered to have more severe climates and are psychologically less invigorating.

## References

Environment Canada. 1980. *Climatic Perspectives* 2(24):3.

Johnstone, K. and P.Y.T. Louie. 1983. *Water Balance Tabulations for Canadian Climate Stations.* Ottawa:Environment Canada, Atmospheric Environment Service, D.S. 8-83.

Phillips, D.W. and R.B. Crowe. 1984. *Climate Severity Index for Canadians.* Ottawa:Environment Canada, Atmospheric Environment Service, CLI-1-84.

# FOUR
# Spatial Growth of Brandon

*John Everitt and Christoph Stadel*

## The Early Development of Brandon: Boomtown and Wheat City, 1881-1914

*The Foundation of Brandon*

The early Canadian settlement of Manitoba was closely associated with initially the fur trade and later the Red River settlers of Lord Selkirk, but neither of these developments had a lasting effect on the landscapes of western Manitoba. The site of the city of Brandon and most of its hinterland did not become part of the province of Manitoba until the provincial boundary expansion in 1881. The enlargement of the "postage stamp province" took in "the territory into which some 16,000 settlers had gone since 1876" (Morton 1967:197) and included the southwest Manitoba region of today. More than just a change in size was involved for, as Morton indicates, "the old Manitoba of 1870 had been engulfed in the new Manitoba of 1881. In one decade of swift change the province had seen the fur trade give way to the grain trade, the cart brigade to the railway train" (Morton 1967:199). Brandon was symbolic of this dramatic change as it was created by the CPR, was sustained by the addition of other rail lines, and soon gained the title of "The Wheat City," a label that it bears to this day.

Brandon's founding was a direct result of the building of the first transcontinental railway, although it was really a sudden climax to the slow growth of the preceding decade. "A Pacific Railway had been promised in 1871. It was ten years before anything much was done about it, but the 'CPR' was to influence the disposition of lands in Western Manitoba long before any track was laid, even before its location had been decided upon" (Tyman 1972:35).

This influence could be seen in many ways, with land speculation and speculative town building being the two most significant results, speculation resulting from the debate over the actual route of the rail line:

> Two possible routes for this railway were under consideration for the push west from Winnipeg. If the original plan for a northern route to Edmonton had been carried out, Brandon might never have come into existence. When the decision to use the southern route was reached, streams of settlers and speculators rushed ahead on foot, by ox cart, and by stern wheeler. (Clark and Stadel 1972:59)

Many came first to the settlement at Grand Valley which prospered for a time, but declined as soon as General Rosser located at the present site of Brandon. The disastrous floods of 1881 and 1882 (Chapter 2) spelled the end for Grand Valley and Brandon became the principal settlement.

Brandon gets its name from the Blue Hills of Brandon located sixteen kilometres to the south (McFadden 1982:18). These had received their name from a Hudson's Bay Company trading post (Brandon House) on the Assiniboine River which in turn was named after Brandon Hill located on an island in the Albany House area of James Bay where Captain James and his crew wintered in 1631. This hill was probably named after a prominent hill located near the centre of Bristol, England (Brown 1982).

Brandon's early history was similar to that of other settlements on the Prairies. Although geographical factors such as the physical environment, distance from other settlements and perceived economic value played a part in its founding, and, although the ingredients for urban growth were present, individual decisions, first by General Rosser and later by administrative officials and residents, were critical in the growth and development of the city.

### *Early Growth: Commercial Development*

Brandon was chosen as a townsite by General Rosser in May 1881 and was laid out by CPR surveyors with streets that were deliberately made narrow to avoid problems that had been encountered in Winnipeg where wide thoroughfares "looked like deserts" when traffic was scarce and where, when wet, the "notorious Winnipeg mud made it a hazardous business to get from one side of the street to the other" (*The Brandon Daily Sun* 18/06/1927). Present traffic volumes have necessitated the introduction of a one-way street system in Brandon's downtown area.

The first passenger train arrived on 11 October 1881 (McFadden 1982:18), speeding up travel to and from the east and putting the steamboats out of business (Chapter 2). The following year, on 30 May, only one year after being chosen as a townsite, Brandon was incorporated as a city of three square miles despite the fact that it was still essentially a "rough, frontier town of tents and shacks that had been erected with stunning suddenness in the prairie wilderness" (Coleman 1957:17-18). Clearly physical conditions in the new city were poor, partly due to the hustle and bustle that resulted from Brandon being not only a growing centre but also "a jumping off place for people going into the surrounding areas and beyond" (McFadden 1982:20).

A Brandon pioneer who remembered this period of "boosterism" and rapid growth recorded that, "It had been discovered that a charter for incorporation as a city would cost the same as that for a town, and therefore, economically, let alone any other reason, it would be wiser to ask incorporation as a city, especially as there was no population requirement, and anyway, sooner or later, a city charter would be required" (*The Brandon Daily Sun* 18/06/1927). Such was the spirit of the times!

Civic offices were built in 1883 and twelve aldermen were elected from four wards (three from each ward), albeit with evidence of the "questionable" election practices that were characteristic of the time (*The Brandon Daily Sun* 18/06/1927). Thomas Mayne Daly, Jr. was elected mayor and later progressed to federal politics. Questionable actions also influenced the location of public buildings; for example, the Court House was located at least two kilometres outside the "city proper," apparently because of "the evil intentions of estate owners and high civic authorities" (*The Commercial* 24/07/1883).

Immediately after the selection of Brandon as a townsite, businessmen and speculators rushed to the site, anxious to gain an early foothold. "Speculators had a field day buying and selling property at constantly rising prices starting the domination of Brandon's early history by the business and professional class — a situation which was to continue" (Clark 1981:1). All the while streets were being laid out by surveyors, and lots were being staked (Coleman 1957:18). Brandon was entering its first phase of major growth and development.

James Smart, a pioneer hardware merchant who was later a mayor of the city, gave the following eyewitness account of

Brandon as it was in June 1881, a month after Rosser's decision was made:

> I well remember my first experience in walking up Rosser Avenue. It was a beautiful clear day. The street was marked out only by the location of three small buildings and one or two shacks facing it. I tramped over the marshy ground west of ninth street through very long grass to reach thirteenth street where a good hotel, the Royal, was being built. The principal business establishments then were a lumber dealer, Railway Contractor, a general store, miscellaneous stores and a livery stable of canvas. There were also two hotels, both of which were canvas. (Quoted in Coleman 1957:19)

The first business in Brandon was reportedly done by Ginnett Bros. in a shanty on the riverfront in July 1881, but the pioneer store was run by Winters and Wastie at the corner of Pacific Avenue and Fourth Street (*The Brandon Daily Sun* 18/06/1927). Rosser Avenue was soon flanked by a plank sidewalk and had become the main business street by 1886 with the "peak value intersection" located at or about Sixth Street and Rosser Avenue (Plate 4:1).

PLATE 4:1

*In 1883 Brandon was just beginning to grow — as shown by this photograph of Rosser Avenue. Streets were unfinished and often mud-filled. A few board sidewalks had been constructed and planks were used to cross ditches and, in some places, the street itself. Most of the buildings were single shacks, but a few larger frame stores and hotels had already been built. The Queen's Hotel was located at the junction of Tenth and Rosser. (Public Archives of Manitoba: Gisli Goodman Collection 30.0)*

The years of rapid growth during the early 1880s were mainly the result of railway expansion to the west, the continentwide economic attraction of the Prairies and the associated steady influx of settlers, but also, in part, a result of the hopes, overconfidence and speculative zeal of the pioneers as McFadden (1982:21) records:

> When W.J. White published the first number of the *Brandon Sun* on 19 January 1882, he had reported a resident population of 700 with a transient population double that and 120 buildings which had cost $200,000. In the spring there was a permanent population of 1,500 with 120 businesses including the Brandon Flouring Mills and the Merchant's Bank; and by midsummer they reported that Brandon was 'a full fledged city with complete municipal machinery, nearly 200 businesses, institutions of every class and a permanent population of not less than 3,000.' There were four banks, a flouring mill, sawmill, two planing mills, several other industrial institutions, two fine new churches and a third under construction, four regular church organizations, numerous imposing business buildings, stocks of goods amounting up to $1,000,000 in value, and hotel accommodations for over 1,000 transient people. There were saddleries, harness makers, blacksmiths, carriage makers, furniture houses, an aerated water works, grocery stores, butcher shops, dry goods stores, farm machinery dealers, jewellers, and many more. There were six hotels in operation and a private club under construction. 'All the north and south streets of the city from First to Eighteenth are being graded and gravelled to the southern limits of the city, while Rosser Avenue and Sixth Street are having cobblestone gutters constructed, and are being coated with screened gravel. Culverts, drains and so forth are being constructed where necessary, so that solid wheeling can at all times be had through the city while ladies can in future walk on clean solid roads half an hour after a thunderstorm.' There was a brick fire hall and council chambers, 5,000-gallon water tanks at different locations throughout the city, nine miles of graded streets, 11,802 feet of eight-foot sidewalks and 11,524 feet of four-board sidewalks. These improvements had cost $85,000.

Despite this boom, there were constant complaints about the shortage of services and of labourers to perform the services. Certainly, the building trade was much in demand as it attempted to eradicate the "scarcity of dwelling houses," the "inadequate school accommodation," and the "totally inadequate" hotel accommodation (*The Commercial* 17/10/1882; 24/10/1882).

In general, Brandon experienced a boom in the early 1880s, but there were constant variations as recorded by *The Commercial*. In November 1882 "business [was] good and promise[d] to keep so" whereas in December "business was somewhat suspended during the stormy weather." By July 1883, "business [had] greatly improved" but later in the same month business was again not "quite so brisk as it was the previous week." In August business was "good" again but in September it was only "fair" although promising "to improve." Clearly life was anything but boring in pioneer Brandon.

Although the boom slackened toward the end of the decade, and although the Prairies experienced a series of economic depressions, Brandon, unlike many other prairie settlements, did not decline. It established a new set of basic public institutions, and its morphology was transformed by a number of impressive brick and stone buildings which replaced the original low wooden structures (Clark and Stadel 1972:61). Brandon, and the surrounding municipalities of Cornwallis, Elton, Daly, Whitehead, and Glenwood, were soon incorporated as the county of Brandon with the city as the judicial centre (Province of Manitoba 1976:1-5), principally because a large courthouse had been built in the city in 1884. However, county government was soon abolished, also in 1884 (McFadden 1982:20).

By 1888 Brandon was becoming a centre for some industry, already having a planing mill, a saw mill, a flour mill (which relocated in Winnipeg during the 1929-35 Depression), and two breweries (which both ceased operation in the early 1970s). In February 1888, the Electric Light Company was established, and the city soon had its first electric lights (Clark 1981:4). Brandon's central position in southwest Manitoba led to its dominance in agriculturally related industries, particularly the thriving horse and grain markets.

The wheat trade boomed because there was little competition for the main CPR line through the city. In the early days wheat was hauled 100-130 kilometres to the Brandon market, to a line of grain elevators along Pacific Avenue, and in 1889 between one and two million bushels of wheat were sold in "The Wheat City" (Plate 4:2). Farmers who came to Brandon to sell their wheat patronized retail outlets which also flourished. By 1890, seven elevators and two flat warehouses with a capacity of over 220,000 bushels existed, making the Wheat City the largest centre, outside Winnipeg, in the province. In addition there was a large flour mill which soon became the second largest outside the capital, after one at Portage la Prairie.

Changing conditions, particularly the growth of new railway lines and companies, soon "robbed Brandon of its distinction as a grain centre" but by this time the city had established its credentials in other areas and could survive the loss. In fact, by the 1890s, Brandon was already "rapidly rounding into shape as a commercial, manufacturing, agricultural, and jobbing centre" (*The Brandon Daily Sun* 18/06/1927).

PLATE 4:2

*Pacific Avenue was well developed by 1885 with a line of grain elevators on its northern side and hotels and stores to the south. The "Wheat City" was the central node for most of southwest Manitoba until railway branchlines were built to serve the outlying areas. Wagons of grain — in bags — were commonly seen in long lines waiting to dump wheat into the elevators. Pacific Avenue was an important commercial thoroughfare at this time, with the Grand View Hotel being located at the corner of Ninth Street and Pacific Avenue. (Public Archives of Manitoba: Brandon 19)*

A setback occurred in 1889 when a disastrous fire destroyed many buildings of wood and canvas in which stores and offices were located. But progress was not slowed greatly and more buildings, now of brick and stone, were built along Rosser Avenue during the early 1890s. "Within a decade of its incorporation on 30 May 1882 Brandon was a city of fine commercial buildings and homes and many of the institutions which were its trademark throughout its first century were already established" (McFadden 1982:21).

In the late 1890s growth was slow in Brandon, which by 1899 had become "a typical Western Canadian market centre-distribution depot" (Clark 1981:8), but this lull was superseded by feverish settlement and building during the first decade of the twentieth century — the period of greatest growth in Brandon's entire history. At this time, the business district spread to the side streets off Rosser Avenue, on to Eighth, Ninth and Tenth Streets. Also, both the

PLATE 4:3

*This view of the Central Business District in 1919 from the northwest shows the skyline dominated by churches, the Prince Edward Hotel (the "Prince Eddy"), and the McKenzie Seeds building. The Great Northern Railway, which owned a branchline (the Brandon, Saskatchewan and Hudson's Bay Railway) that ran from North Dakota to Brandon, had a station located in the middle foreground. This fine brick building cost $37,655.26 to build in 1906 and was once intended as a joint terminus for the Great Northern Railway and the Grand Trunk Pacific Railway. The branchline closed in 1936. (Public Archives of Manitoba: Brandon 5)*

Canadian Northern Railway and the Great Northern Railway reached Brandon during this period. A new Brandon Hospital for Mental Diseases was built to replace one that burned in 1910, and in 1911 the Prince Edward Hotel, a major and imposing structure in the cityscape, was constructed (Plate 4:3). In addition, between 1910 and 1914 the Brandon waterworks system was expanded (Chapter 2), streets were paved, streetcars were introduced (1913) and central steam heat was provided, along with gas and electric power, for industry (Province of Manitoba 1976:1-5).

In general, the period between the turn of the century and World War I was one of mostly rapid, but not completely steady, growth, a pattern reflected by the value of building permits issued (Table 4:1). Growth was initiated by a series of good agricultural years, underlining the close links between Brandon's prosperity and the agricultural economy of its hinterland, and resulted in Brandon's population tripling in the period from 1900 to 1914, reaching fifteen thousand (Figure 4:1). About half the city's built-up area dates from the years preceding World War I (Figure 4:2), the relatively unchanged commercial downtown district and the residential areas immediately surrounding it still providing evidence of this early growth period. The commercial district expanded continuously, but

TABLE 4:1
VALUE OF BUILDING PERMITS IN BRANDON: 1904-1944

| Date | Value in $ | Date | Value in $ | Date | Value in $ |
|---|---|---|---|---|---|
| 1904 | 423,173 | 1918 | 99,022 | 1932 | 33,088 |
| 1905 | 434,550 | 1919 | 98,541 | 1933 | 46,821 |
| 1906 | 654,902 | 1920 | 412,829 | 1934 | 44,758 |
| 1907 | 541,281 | 1921 | 741,190 | 1935 | 113,235 |
| 1908 | 292,869 | 1922 | 225,029 | 1936 | 56,310 |
| 1909 | 349,680 | 1923 | 183,034 | 1937 | no data |
| 1910 | 940,128 | 1924 | 270,579 | 1938 | 50,085 |
| 1911 | 1,024,529 | 1925 | 76,579 | 1939 | 83,695 |
| 1912 | 1,166,214 | 1926 | 226,138 | 1940 | 360,446 |
| 1913 | 609,609 | 1927 | 230,252 | 1941 | 258,648 |
| 1914 | 395,335 | 1928 | 428,138 | 1942 | 271,120 |
| 1915 | 36,155 | 1929 | 403,667 | 1943 | 189,359 |
| 1916 | 244,873 | 1930 | 197,245 | 1944 | 400,675 |
| 1917 | 180,338 | 1931 | 286,611 | | |

*Source:* Brandon Building Permits Register (1904-1918) and Henderson Directories.

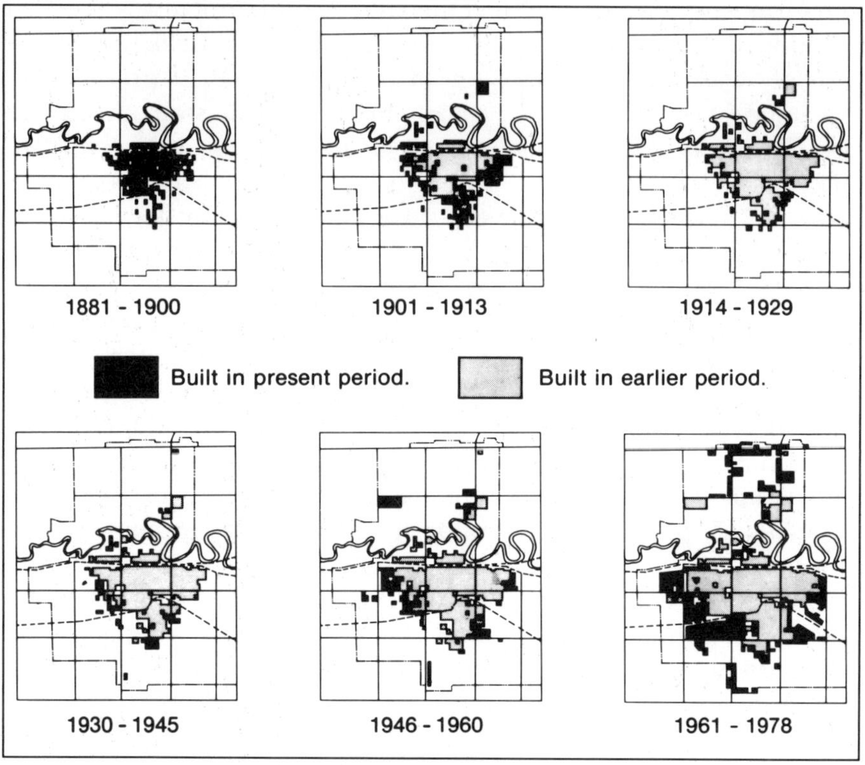


FIGURE 4:1
*Brandon's population growth: 1881-1978*

it was still overwhelmingly oriented towards the core of the city (Plate 4:4).

The mix of businesses changed over the years as the city stabilized and matured and as the urban area and its populace reflected changes in Canadian society in general. In 1883 Brandon's occupational structure had been dominated by building trades and services which were trying to cope with, and indeed to fuel, the city's growth — merchants, the clothing trades and other services. By 1894 only minor changes had occurred, but by 1914 major differences could be seen. The traditional building trades, merchants, clothing trades and service trades remained important but the largest number of jobs was now in financial services — banks, insurance companies, loan companies and related businesses — reflecting the maturity of the city's economic structure, and indicating the recognition by outside, commonly eastern, entre-

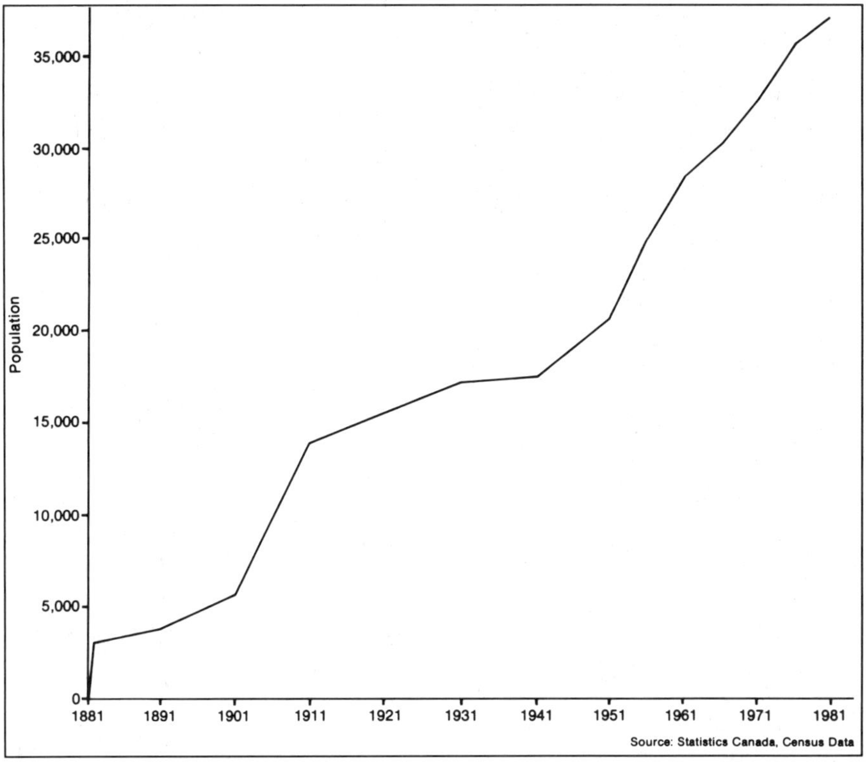


FIGURE 4:2
*Spatial growth of the built-up area of Brandon: 1881-1981*

preneurs that the growth of Brandon would continue. Also, there were additions in professional services, in the communications sector, and a notable rise in metal work trades as the automobile became popular. But not all the changes were in the direction of growth and expansion. Indeed, as World War I approached, problems that reflected structural difficulties within the city and its region became apparent. This change in "circumstances" may have been substantial by as early as 1914:

> As a result of what many deemed to be the highest taxes in Western Canada, tax arrears were escalatory and, consequently, civic revenues were declining. Furthermore, and to the disappointment of many, Brandon's newly-inaugurated street railway system was losing money at a projected rate of $1000 per car per year. While the 300 civic employees who had their wages reduced in

mid-1913 were among the first to suffer, the impact soon became widespread. Brandon, as a distribution centre and as a market town, was heavily dependent upon its rural hinterland and that area, due to the decline in immigration and the advanced degree of development which had already occurred no longer exhibited the same heavy thrust for consumer goods that it once had. Although both the CPR and the Roblin government of Manitoba attempted to alleviate a growing unemployment problem in 1914-15 by commencing construction work in advance of the scheduled date, the local economy had clearly soured by the outbreak of war. (Clark 1981:69)

PLATE 4:4

*A view taken in 1919 of the 700 block of Rosser Avenue, looking west. The size of the buildings indicates the considerable growth and development of the commercial sector of the city during the early years of the twentieth century. The Empire Hotel (727-773 Rosser Avenue) was run by A.E. Evans and had the Woolworth Department Store beneath it. A mixture of horse-carts and automobiles is evident and streetcar tracks can be seen in the street. (Public Archives of Manitoba: Brandon Streets — Rosser 3)*

*Early Growth: Residential Development*

Rapid early growth and intense land speculation during the late nineteenth century were characteristics of many urban settlements on the Prairies, ranging in size from villages such as Rapid City to major cities such as Winnipeg (Artibise 1981). Brandon fit this pattern, its designation as a divisional point on the main CPR transcontinental line giving it an added impetus. Its population increased elevenfold between 1881 and 1901, and nearly tripled again by the World War I (Figure 4:1). By 1914 Brandon had established itself as the seventh largest urban centre on the Prairies, and the second largest in Manitoba.

The years between 1881 and 1914 were a boom period for Brandon's spatial growth, as well as a period of population growth. The growth was encouraged by the arrival of the Canadian Northern Railway in 1900, which took over the Northern Pacific interests, and by the Great Northern Railway in 1906, which entered the city as the Brandon, Saskatchewan and Hudson's Bay Railway. The third Canadian transcontinental, the Grand Trunk Pacific, passed just north of Brandon, and for local political reasons never entered the city. It was later amalgamated with the Canadian Northern Railway to form the Canadian National Railway (CNR).

Initially, residential development was near the commercial core, although it was of a temporary and sporadic nature consisting mainly of tents, hotels and boarding houses, the latter socially segregated by occupational status. This arrangement reflected in part the uncertainty of the settlers about the future, in part the difficulty of travelling even short distances over poor or nonexistent streets and in part the fact that the male-dominated society did not need dwellings outside the core. It was also partially a result of the shortage of building materials and of skilled labourers who could build permanent homes. Also, businessmen often used backrooms or the upper floors of their working places as living quarters. By 1883, however, the pattern of growth was clear, with a few houses for wealthy people being built one or two kilometres from the commercial core (Figure 4:2). However, the spread of settlement was not continuous and infilling did not take place for many years (Westenberger et al. 1985).

Growth was mainly to the south, in a semicircle around the core, expansion to the north being restricted by the barriers of the CPR mainline and the Assiniboine floodplain (Figure 4:3). Most of this

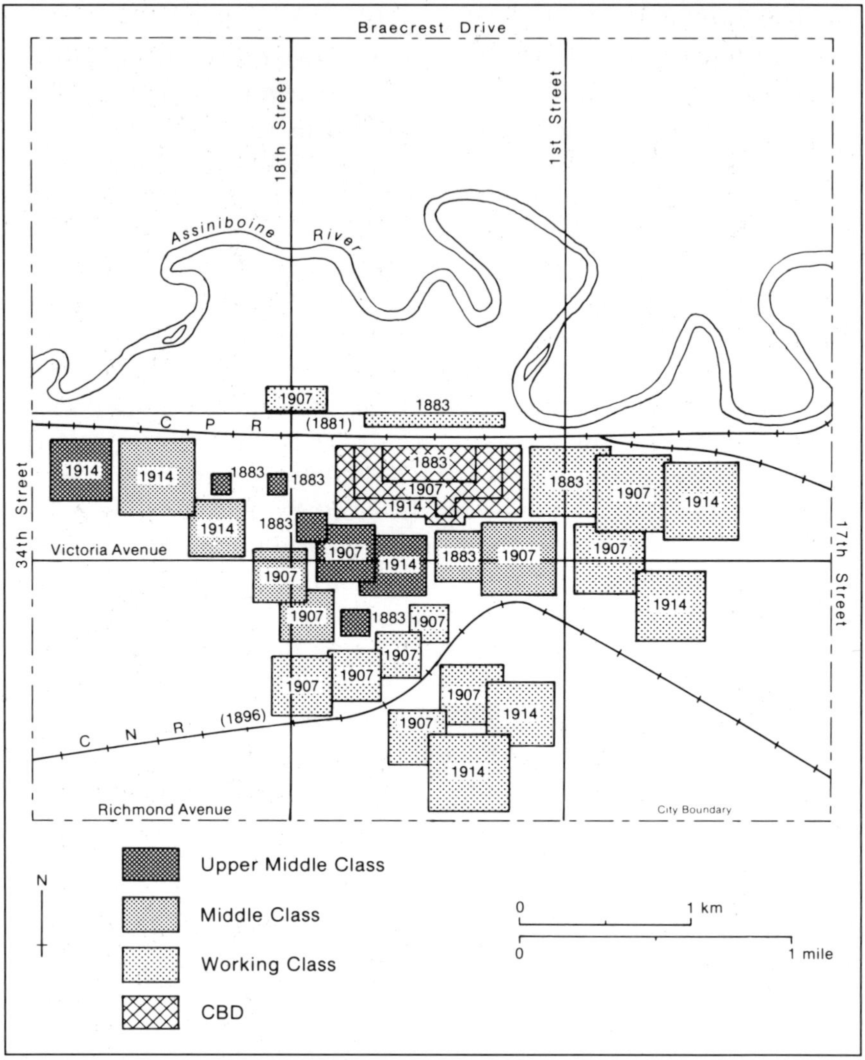


FIGURE 4:3
*A model of Brandon's social areas: 1883-1914*

expansion consisted of socially mixed residential land use; commercial and industrial activities were largely confined to the downtown core which did not extend much beyond the area around the CPR station.

Although by the end of the 1880s the city owned both the river bridges, at Eighteenth Street and First Street (Chapter 2), so that people could cross the river without charge, little development occurred north of the river. But there was speculative buying on the North Hill as early as 1882, by entrepreneurs who hoped "to make this the aristocratic part of town." Although "the situation certainly is good, having a beautiful slope towards the south" (*The Commercial* 21/11/1882), the barrier effect of the river and the railway was too great for most people and urban growth continued to take place mainly south of the river. Although the influence of this permeable barrier is now decreasing, it has had a marked effect on the configuration of the urban area.

Although permanent housing was sparse in 1883 a spatial pattern that is still in evidence was beginning to emerge (Figure 4:3). The upper-middle class (using terminology defined in Westenberger et al. 1985) was located predominantly on the southwest side of the downtown area. That part of the middle class that did not reside in the downtown boarding houses lived mainly in the south and west including a small cluster of middle-class houses between Seventh and Tenth Streets. Working-class areas were developing on the east side of downtown, and north of the CPR tracks.

Three reasons for this trend-setting pattern can be hypothesized. First, the physical environment of the floodplain and the existence of the CPR tracks made the north side of the city difficult to reach and undesirable to live in. Second, the prevailing winds from the west (Chapter 3) and the growth of the railway yards meant that the "east end" was more noisy and noxious than the south and west sides. Third, once this pattern of development had been established, the higher-class groups moved to the south and west to avoid the people as well as the environment on the other side of the business core.

By 1907 considerable changes had taken place, both in the area covered and in the pattern of development. The downtown area had grown, both in size and significance, and had become better defined as a commercial core (Figure 4:3). The initial speculative boom had subsided and Brandon residents were now interested in establishing a permanent home, resulting in growth of the residential areas, and a decrease in temporary dwellings close to downtown. Housing had spread southwards to the one hundred-foot wide Victoria Avenue, along which shade trees had been planted and many "grand" houses constructed (Barker 1977:28).

The industrial sector of the city was becoming better developed and was exerting a stronger influence on the residential pattern, rail line developments being particularly influential. During the late nineteenth century the Northern Pacific tracks entered the city from the southeast, crossing diagonally the grid patterned roads to reach a terminus close to downtown at Ninth Street and Princess Avenue. Then in 1903 the CPR yards were enlarged and relocated east of First Street, and later, in 1906, the Canadian Northern built a track through the southern part of the city to a new set of terminal facilities straddling First Street at McTavish Avenue. "South end" citizens "grew vociferous" concerning this latter move but to no avail (Barker 1977:92 and 106). It was clear that "railways meant importance and success. Brandon's future depended upon the number of steel highways touching it," and the city council believed that the city "must become an undeniable ironhorse hub" (Barker 1977:99).

The working-class zones grew most, so that by 1907 they made up the largest area of any group (Figure 4:3). Some expansion took place adjacent to the core, in neighbourhoods previously occupied by middle and upper-middle-class groups, but most growth was in the south and east ends of the city. This was in part a result of the general broadening of Brandon's economic base, but reflected particularly the construction of the Northern Pacific and Canadian Northern trackage in the south and the expansion of the CPR yards in the east. Ethnically the expansion was significantly Eastern European, mainly by people of Polish and Ukrainian origin.

The Canadian Northern lines also blocked the expansion of upper-middle-class housing to the south, although the earlier elite sector in the west and southwest continued to expand. In fact, the upper-middle-class social area consolidated to the west of Tenth Street between 1883 and 1907, although it was surrounded by a middle-class zone which, by 1907, was almost completely cut in half by the expansion of the commercial core south along Tenth Street, by then the major north-south artery.

The growing significance of the western part of the city was demonstrated by the creation, by real estate promoters in 1905, of Brandon's first "suburb" in the west end (Barker 1977:94). East of the Tenth Street artery, the middle-class zone was more uniformly constituted and more compact than middle-class areas to the west. However, it was not significantly larger than in the 1880s although it had spread southeastwards towards the intersection of First Street and Victoria Avenue. Expansion, both to the east and the west, demonstrated typical sector development with outward movement of the elite from the centre of the city.

By 1914 there had been substantial areal expansion (Figure 4:2) but little locational change in social areas (Figure 4:3). The commercial core continued to expand, particularly along the east-west axis of Rosser Avenue and the north-south artery of Tenth Street, a pattern that was reinforced by the short-lived introduction, in 1913, of streetcars which plied these major thoroughfares. The upper-middle-class neighbourhoods had increased in extent and had consolidated noticeably southwest of the core, whereas the middle-class zones remained quite similar, in total area, to the situation that existed in 1907. However, the amount of middle-class housing diminished east of the downtown commercial area where working-class numbers had increased. Also, a large middle-class area had become evident west of Eighteenth Street, a directional bias which has continued ever since.

As might be expected, the working-class districts continued to be the largest in the city with pockets of working-class residential development near the core, and with three almost uniformly working-class areas east of First Street, north of the CPR tracks, and in the "south end" of the city. In all three cases railroad development, and consequently employment, was the critical factor and in all cases the working-class had filled spaces left unused by the other two social groups (middle class and upper-middle class). East of Tenth Street and south of Victoria Avenue the growth of middle-class residential zones was truncated by the active expansion of working-class developments. Some neighbourhoods were abandoned by the middle classes and expansion of higher-class housing was "deflected" into the west end where such social barriers did not exist.

It should be emphasized that the construction of buildings was discontinuous and that infilling took place for decades as on Sixteenth Street, south of Victoria Avenue, where the first dwelling was built in 1888 and the last in 1948, with a sporadic pattern of construction between these two dates.

A number of major social and economic forces can be identified that help to explain the development of the city's social areas. First, the upper-middle classes had greater flexibility, freedom of choice, and power in determining residential location than the lower classes; they therefore picked the choice locations. Second, although there are no major relief variations in Brandon, local relief, drainage, and microclimate may have played some role, at least as a reason for the elite avoiding northward expansion on to the floodplain of the Assiniboine River, and eastward expansion into areas affected by air and noise pollution from downtown and the railroad yards.

Third, once the upper-middle class had selected its locations, the middle class located close by, a strategy identified elsewhere as a means of minimizing social distance between individuals and the groups they wish to emulate (Timms 1975:98). As a result, members of the working class had little choice although, as in other cities, a tendency for extremes of the social scale to locate at opposite sides of the city is evident (Duncan and Duncan 1955:493). In Brandon a relative lack of mobility and the specific location of economic opportunities reinforced this pattern. There is also evidence within the working class of a spatial separation between skilled and unskilled workers, the latter being found more commonly north of the CPR tracks on, or adjacent to, the floodplain. Although by 1914 this area was ethnically dominated by Eastern European immigrants, the overall number of non-British in Brandon was not large.

It is also worth looking at Brandon's spatial development in the context of traditional models of urban morphology, although the physical barriers of the river and the floodplain, and the physical and social barrier of the CPR tracks, mean that models must be cautiously applied. Figure 4:3 shows the major features which shaped the spatial growth and social differentiation of the city; that is, the Assiniboine River and floodplain, the city boundary, and the railways. The model shows in a generalized fashion the truncated areal expansion of the Central Business District (CBD) and the shifts of residential areas occupied by the three major social groups. It was built using a combination of individual housing-unit data collected by Pirker (published in Everitt and Stadel 1983) including the age of each residential dwelling in the city, and the social area data for 1883, 1907 and 1914 (Westenberger et al. 1985). The model enabled us to compare Brandon and other urban areas for which similar research has been conducted.

In 1883, the distribution of social areas around the commercial core was zonal, sectoral, and nucleated with the upper-middle and middle classes on the south, east and west sides and the working classes on the east and north sides. The arrival of the Northern

Pacific and the Canadian Northern Railways disrupted this pattern, stimulated new land use nuclei, and attracted working-class employment and residences to the south end of the city. By 1907, the working-class areas were largely peripherally located surrounding the residential areas of the upper-middle and middle classes, almost in a semicircle. By 1914, an increase in numbers in the working class had resulted in an enlarged working-class area; the other two groups exhibit no substantial change.

In Brandon, as elsewhere, social areas assume zonal, sectoral, and nucleated patterns, but factors peculiar to the city have to be understood; the influence of the river, the railways, the streetcar system, the road system, the early land developers, and the evolving land use pattern have all to be considered. These environmental and social factors affected the decision making of early builders and, in some cases, the results of the decisions can be seen to this day.

## The Emergence and Consolidation of Brandon as a Regional Centre, 1914-1980

### *Growth in the Interwar Period*

Between 1914 and 1945 there was little increase in Brandon's built-up area but in-filling continued on lots that had often been held by speculators (Figure 4:2). The war years, 1914-18, and the years of the Great Depression saw fluctuations in Brandon's fortunes (Table 4:1). Generally World War I is blamed for bringing an abrupt end to optimism and to the resulting boom. The years after the war were characterized by a series of strikes, and a desperate agricultural situation resulting from drought and general economic depression. It was a period of uncertainty and stagnation although, once more, Brandon, unlike many other small prairie centres, did not decline (Clark and Stadel 1972:61). Both labour and farmers emerged as organized political forces on the local scene, although the established parties remained regionally dominant (Clark 1981:114).

The post-World War I years saw some physical changes in Brandon's structure but the pace of growth was diminished. For example, the opening of the F.W. Woolworth Company store, the city's first chain store (Plate 4:4), signalled a significant retailing future for Brandon, and the arrival of larger numbers of automobiles foretold some of the spatial changes which would soon occur. The decade 1919-29 was especially important in this respect; in 1922 approximately 1,400 vehicles were registered in Brandon; by 1927 this figure had risen to 2,864. The automobile had a great impact on lifestyles and changed the appearance of the city. Most importantly, automobile travel expanded the city's trading area and let it recapture its former dominance in southwest Manitoba.

The automobile had a negative effect on the troubled streetcar system. This had been established only in May 1913 and did quite well, financially, during its first years; later, deficits led to criticism and calls for its closure. The system was never expanded beyond its initial ten kilometres of track, centred on the CBD and the older housing areas. In the early 1930s the streetcars were slowly abandoned and on 30 April 1932 ceased to run. They were eventually replaced by a succession of equally contentious bus systems.

Reflecting the local economy, the population of Brandon grew slowly but steadily (Figure 4:1). The boom years of 1901-11, when the population total increased by 146 percent, were clearly over. Between 1921 and 1931, the population grew by 10.9 percent, perhaps boosted by returnees from the war, but in the next decade growth was only 1.8 percent as Brandon experienced the Great Depression. After a few years of the Depression, rising relief costs and declining tax revenues created a municipal deficit. By 1935, the city "was forced to consider the possibilities of bankruptcy, repudiation of its debts and the surrendering of financial control to the provincial government" (Clark 1981:156). The financial crisis reached a climax early in 1936 when the provincial government appointed a financial advisor for Brandon, although the mayor and council continued in office with limited spending powers (Clark 1981:136). One result was a halt in construction. Later, during World War II, there was a shortage of materials for nonmilitary purposes. However, the city came out of the Depression and World War II in good shape, although economic recovery was neither immediate nor uniform. There are two reasons why the city emerged in a good position: in 1938 wheat prices rose significantly, giving the "Wheat City" a real economic boost and during the war Brandon was a major military training depot for dominion forces, and its business reflected the proximity of Camp Shilo and an air station at Rivers (Province of Manitoba 1976:1-6).

*Growth in the Post-World War II Period*

The post-World War II years brought the third period of growth to Brandon. "The city enjoyed a small financial surplus in 1946; at least a dozen new industries began production in these 12 months while Brandon retailers reportedly enjoyed their greatest year ever in 1947. This new found prosperity, however, was not yet evenly distributed" (Clark 1981:196-97). Even city services were not universally available; for example, Brandon's sewerage system had not been extended north of the tracks, those who had chosen to live in the area being warned that extending the system into their neighbourhood might never be economically feasible because of the inadequate gradient.

For most people, however, the times were good:

> Hiring conditions were slowly improving and future prospects were encouraging. . . . The local housing shortage appeared to have lessened; Brandon's new and pre-paid water treatment facility was in operation; the city's capital debt had been reduced; this western Manitoba railway and agricultural market centre now even had airplane connections with the outside world, thanks to its new municipal airport and TCA [Trans Canada Airways]. (Clark 1981:197)

While the population grew by only 3,500 between 1911 and 1941 it then increased quite rapidly — to 28,166 by 1961 and to 36,242 by 1981 (Figure 4:1). This period of a major population growth was reflected by changes throughout the urban area, including dramatic alterations in the commercial and industrial patterns (Chapter 6) as well as in recreation and transportation. Politically, the traditional pattern was maintained with the business and professional class remaining firmly in control (Clark 1981:224).

Of particular note, because it had great areal implications, was the alteration in the residential base of the city. Building starts showed an oscillating pattern, as they had done throughout the city's history, peaking in the mid-1950s and mid-1970s. "The 1950s were the period during which the City grew mostly to the west towards 34th Street as well as spreading southward. School and other community facilities were also improved and expanded" (Province of Manitoba 1976:1-6). This trend was continued during the 1960s and 1970s (Figure 4:2) with the development of modern residential suburbs (Figure 4:4) which proved to be socially desirable to new home buyers (Everitt and Stadel 1980). The new residential areas were built in a pattern of bays, crescents and cul-de-sacs in marked contrast to the old grid pattern. Also, for the first time there was significant, mostly commercial, development to the

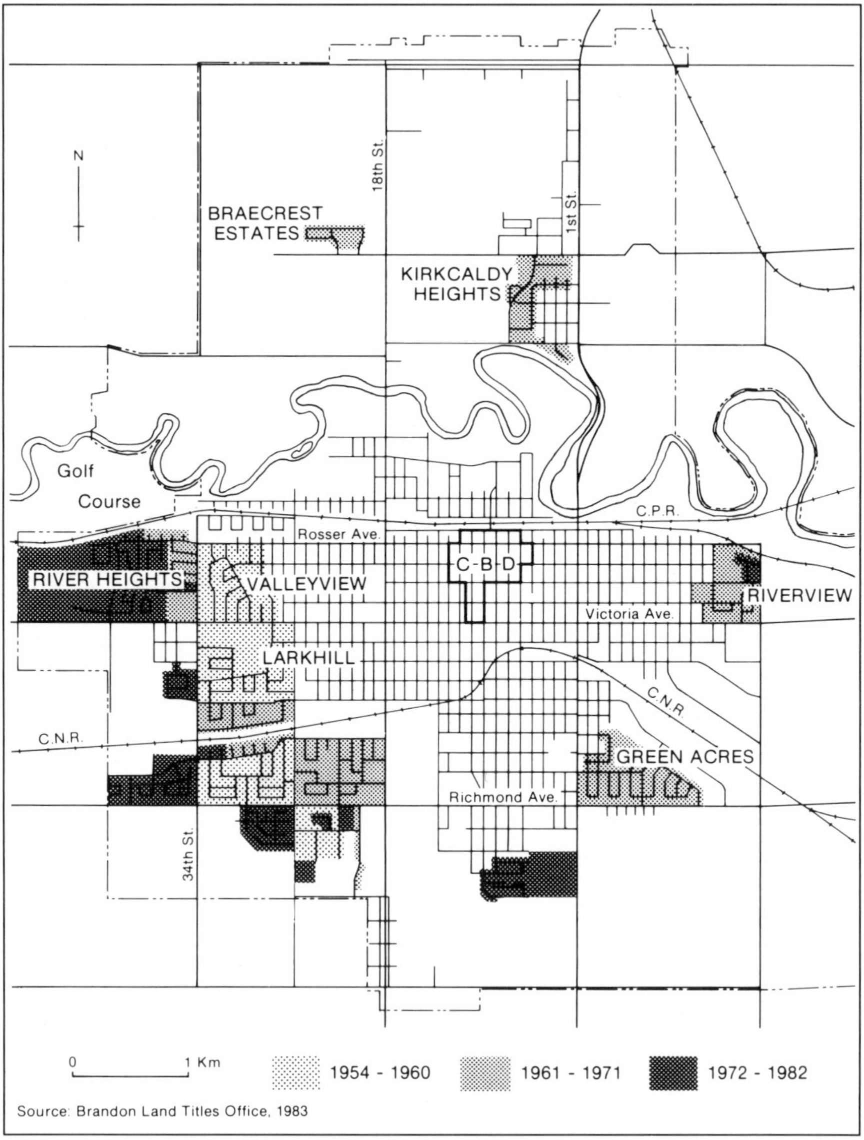


FIGURE 4:4
*Residential subdivisions of Brandon: 1954-1982*

north, particularly along First and Eighteenth Streets and along the Trans-Canada Highway By-Pass which was opened in 1958 (Figure 4:2). Suburban residential growth during the early and mid-1980s has continued towards the west onto land which was relatively cheap to develop, leaving large open tracts of land within the city, most noticeably along the northern fringe (Everitt and Stadel 1980).

Land prices in the residential suburbs have increased steadily since the major phase of suburban development began, although a 1975 comparison between Brandon and Winnipeg residential lot prices showed Winnipeg's prices to be almost 50 percent higher. The difference can be accounted for by higher raw land prices and higher developer profits in Winnipeg, and lower servicing costs in Brandon (Province of Manitoba 1976:2-17). Despite this, spiralling holding costs, such as interest on loans and property taxes, coupled with stagnant land prices in Brandon, have contributed to the demise of some building companies. Apart from land prices and climbing interest rates, the major reason for the increase in housing prices has been increased building costs. Between 1975 and 1982, residential construction costs in Brandon rose by nearly 80 percent from $27 to $48 per square foot (Province of Manitoba 1976:2-17; and authors' field survey, 1982).

During the late 1950s and early 1960s, many Brandon residents began to trade in their old homes in favour of newer, single-family homes (*Brandon Sun* 01/08/1962). This resulted in westward intraurban movement into new suburban residential developments such as the Valleyview and Larkhill Subdivisions (Figure 4:4). One result was that by the 1960s Brandon was faced with a shortage of low-income housing, as the builders concentrated on housing for middle- and high-income families. This was accompanied by declining housing quality in the older residential sections and eventually caused modest development of low-income housing, such as Argyle Courts, built in 1968-69 in what was then the southwest corner of the city.

After a lull in the late 1960s, the 1970s saw a resurgence in single-family housing starts which was also associated with intraurban migration, as 50 to 60 percent of the home buyers were from older areas of Brandon, while the rest were new immigrants to the city. The shortage of housing for low-income families and senior citizens became even more acute.

Movement to the suburbs has been accompanied by obsolescence of the housing stock in some of the older residential areas. Also, new shopping areas have sprung up in an outer belt around the city, associated with the decline of the commercial and social dominance of the CBD, a situation which is common throughout North America. Partly to counter this decline, a new downtown mall was opened and Brandon is participating in a province-wide effort to revitalize and "spruce-up" the core area (Chapter 7).

## A Conceptual Model of Suburban Development in Brandon

The conceptual model (Figure 4:5) is an attempt to portray the factors affecting suburban development in Brandon and to show the major chronological phases in the growth of the suburbs. The location and expansion of suburban residential divisions is shown in relation to the CBD, the arc of older residential neighbourhoods surrounding the CBD, major land use features and transportation lines that affected suburban growth, the floodplain of the Assiniboine River with its adjacent bluffs, and the city's boundaries. Some of these features acted as stimuli for suburban development whereas others were constraints.

While the model is useful in showing spatial factors which influenced the suburban growth process, it cannot show nonspatial factors such as the influence of individuals on suburban expansion, or the effect of socioeconomic differences in the residents.

The uneven spread of Brandon's suburbs is the result of a complicated mix of environmental, social, and political forces. In some cases spatial trends which emerged decades ago have continued by historical inertia, whereas in others the influence of real estate companies and developers, of planners, and of municipal and provincial governments has been paramount.

The fast expansion of residential suburbs towards the western and southwestern edge of the city follows the prewar preference of the middle and upper classes. This pattern accords with the findings for other North American cities that a city generally expands fastest in the direction followed by the upper and middle classes. Growth to the west is also a response to the establishment in the early phases of the city's history of railway-oriented activities east and south of the original core, and to the postwar location of major industries just outside the pre-1972 eastern boundary. Although during the 1950s and 1960s residential suburbs filled a large area

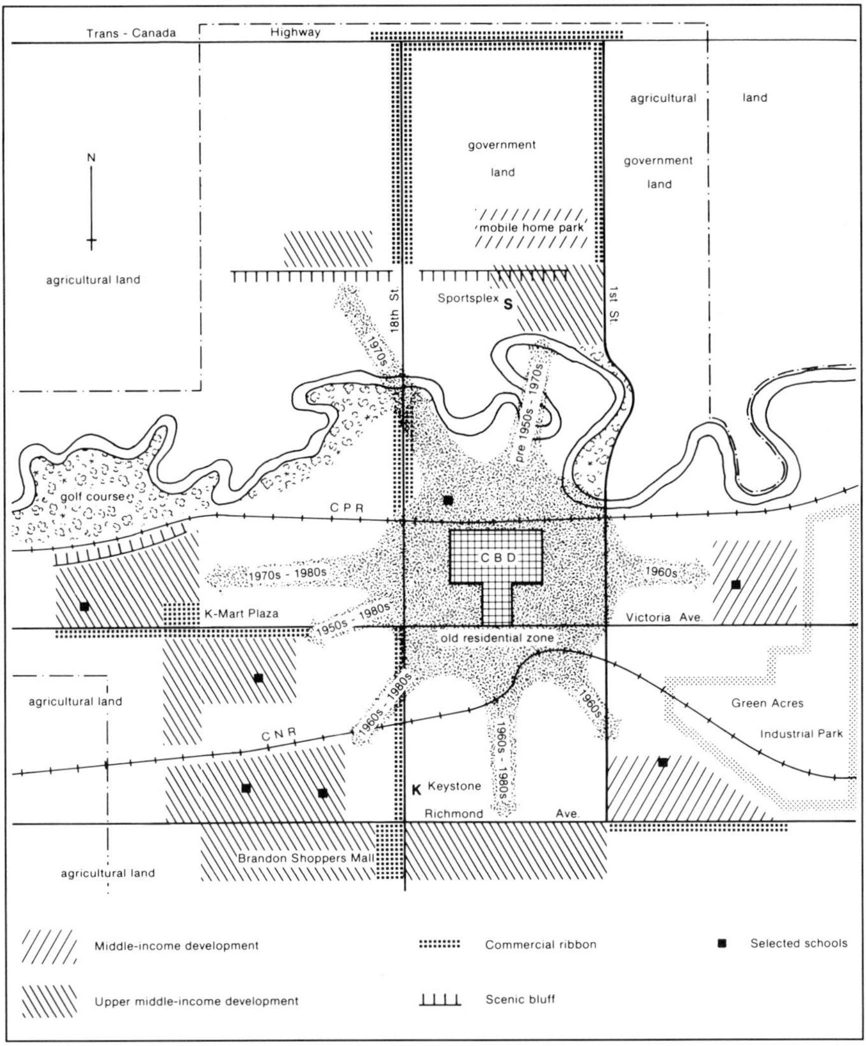


FIGURE 4:5
*A conceptual model of suburban development in Brandon*

east of the prewar commercial and residential core, further eastward expansion was limited by a large tract of land that was zoned industrial in the 1960s, and by the railway yards of the CNR spurline.

Expansion to the north was hampered by a triple barrier: the CPR mainline; the working-class ethnic neighbourhood "north of the tracks"; and the flood-prone valley floor of the Assiniboine River (Chapter 2). The effectiveness of these barriers has diminished in recent years with the improvement of roads and bridges which cross the railway, weakening the social barrier effect of the upgraded neighbourhood "north of the tracks." Also, residential development on the south-facing bluffs of the Assiniboine Valley (North Hill), and to a lesser degree on the edge of the floodplain, has been stimulated by "discovery" of the scenic qualities of the "Hill" and of the waterfront, by the release of government land for residential use, and by the construction of the Sports Complex (Sportsplex) for the Canada Winter Games in 1979. The construction of a new school north of the river has also boosted the fortunes of this area.

In contrast, the 1972 movement of the city boundary north to the Trans-Canada Highway By-Pass did not result in residential expansion beyond the old boundary, although a large portion of this land is now owned by the Canada Mortgage and Housing Corporation (CMHC) and is scheduled for residential development in the second stage (after 1986) of the *Brandon Development Plan* (Province of Manitoba 1977).

Residential growth to the south may have been hampered initially by the CNR spurline and by the Keystone Exhibition Grounds; nevertheless, it quickly reached the southern boundary of the city. Three events spurred development at the southern edge during the 1970s: the 1972 extension of Brandon's boundary; the emergence of a major commercial core with the building of the Brandon Shoppers' Mall at Eighteenth Street and Richmond Avenue; and the building of Crocus Plains Regional Secondary School west of First Street, south of Victoria Avenue.

Given the constraints for residential development at the eastern, southern and northern edges of the city, it could be argued that the rapid and large expansion to the west and southwest occurred "by default." However, real estate companies and developers, aided by city politicians, quickly realized the general spatial trends of urban growth, and recognized the profit potential in acquiring large tracts of agricultural land for resale later and in the promotion and development of suburbs for the middle and upper classes.

The westward spread was assisted by a number of factors: the attractive physical layouts of subdivisions; the social status that was quickly given to the new developments; an expanding commercial ribbon along the major western highway access to the city (Victoria Avenue — Highway 1A); the extension of the city boundary in 1972; the building of a new, attractive school in the Riverheights subdivision; and favourable, or at least not adverse, topographic and landscape conditions. It appears that the growing distance to the core, the close proximity of the CPR line, the open, monotonous, windswept character of the land and the high cost of land taxes are not sufficient deterrents to stop this area from continuing to be the preferred suburban location for Brandon residents.

In terms of socioeconomic characteristics, the differences between the various suburbs appear to be less marked and rigid than in many larger cities. Indeed many residential subdivisions exhibit a variety of social classes, reflected in the variable quality of individual residences. This is particularly true in the case of Kirkcaldy Heights on the North Hill and of the developments on the southern edge of the city. The suburbs near the eastern and western edges of Brandon appear to be the most uniform — predominantly lower-middle class in the "East End" and mostly middle and upper-middle class in the "West End."

As far as spatial dynamism and growth are concerned, development in the east end was generally completed by the mid-1960s, whereas development at the southern edge of the city was particularly active in the 1970s, slowing in the early 1980s. On the North Hill an expansion of the residential subdivision is still in progress and is likely to accelerate in the future. At the western periphery, the growth of suburbs has been by far the most pronounced, especially since the mid-1970s.

## References

Artibise, A.F.J. 1981. Boosterism and the development of prairie cities, 1871-1913. In A.F.J. Artibise, ed., *Town and City: Aspects of Western Canadian Urban Development.* Regina: Canadian Plains Research Center.

Barker, G.F. 1977. *Brandon: A City 1881-1961.* Altona: Friesen and Sons.

*The Brandon Daily Sun*, 18 June 1927.

*Brandon Sun*, 1 August 1962.

Brown, R. 1982. *The Brandon Hill Connection.* Brandon: Tourism Unlimited.

Clark, W.L. 1981. *Brandon's Politics and Politicians.* Brandon: *Brandon Sun.*

Clark, L. and C. Stadel. 1972. Brandon. In *Southern Prairies Field Excursion Tour Guide*. Regina: University of Saskatchewan, Regina Campus.

Coleman, M. 1957. *The Face of Yesterday: The Story of Brandon, Manitoba (1883-1956)*. Brandon: Leach.

*The Commercial*, 17 October 1882, 24 October 1882, 21 November 1882, 24 July 1883.

Duncan, D. and B. Duncan. 1955. Residential distribution and occupational stratification. *American Journal of Sociology* 60:493-503.

Everitt, J.C. and C. Stadel. 1980. A study of power and politics in the urban fringe of Brandon, Manitoba. In A.A. Seaborne, ed., *Some Current Research on the Canadian Plains*. Regina: University of Regina.

Everitt, J.C. and C. Stadel. 1983. Spatial dimensions of the urban growth of Brandon, Manitoba, 1882-1982. In *Background Readings for the Geography of Manitoba, No. 2*. Grand Forks, North Dakota: University of North Dakota, Department of Geography.

McFadden, E.M. 1982. Instant city: the birth of Brandon, 1882. *The Beaver* Summer: 14-21.

Morton, W.L. 1967. *Manitoba: A History*. Toronto: University of Toronto Press.

Province of Manitoba. 1976. *Background Studies: Brandon Development Plan*. Brandon: Province of Manitoba, Municipal Planning Branch.

Province of Manitoba. 1977. *Brandon Development Plan*. Brandon: Province of Manitoba, Municipal Planning Branch.

Timms, D.W.G. 1975. *The Urban Mosaic*. Cambridge: Cambridge University Press.

Tyman, J.L. 1972. *By Section, Township, and Range: Studies in Prairie Settlements*. Brandon: Assiniboine Historical Society.

Westenberger, M., J.C. Everitt and C. Stadel. 1985. The development of Brandon's social areas 1881-1914. *The Albertan Geographer* 21:79-95.

# FIVE
# Population Structure and Social Areas

*Alison Gill*

## Introduction

In large cities, the distinctiveness of different neighbourhoods is evident. This chapter addresses the question of whether or not smaller communities, such as Brandon, with less diversity of function, exhibit similar distinctiveness in terms of social area. Further, if this is the case, are criteria similar to those used in the definition of social areas in large cities applicable? Factorial ecology techniques, which use census data to investigate the social character of urban areas, are widely applied. Several studies of Canadian cities have used this approach, although to permit comparison with the Brandon study, attention is focussed on those of western Canadian cities.

Davies (1975), in his study of Calgary, suggests that six factors account for two-thirds of the variance in community areas, and that of these, four main ones account for the greatest proportion of variance in the data set. These four sources of explanation correspond to family status, economic status, migrant status, and ethnic status. These findings are similar to those from other cities such as Edmonton (Davies 1978), Winnipeg (Nicholson and Yeates 1979) and Vancouver (Peucker and Rase 1971). The spatial correspondence of family status, economic status, and ethnic status with concentric, sectoral, and nucleated patterns respectively, is also commonly exhibited, although unique features of individual cities result in some distortions to the basic patterns.

The study of Brandon's social areas was based on 1971 census data for enumeration areas within the legal city limits of Brandon (Dominion Bureau of Statistics 1971). In 1981, Brandon had a population of 36,245, which represents a modest growth of 16 percent over the 1971 population of 31,150. Although population is relatively stable, two notable changes in the population structure occurred during this period. Most evident is the increase in the proportion of both male and female residents in the 25-34 year age group (Figure 5:1). A possible explanation for this may be found by referring to the difference shown between the 1971 population structure of Brandon and that of Winnipeg (Figure 5:2). In 1971, a major difference in the population structure of the two cities was the lower proportion of Brandon residents in the 20-30 age group. This probably reflected the tendency of young people to leave Brandon for better employment and higher education opportunities

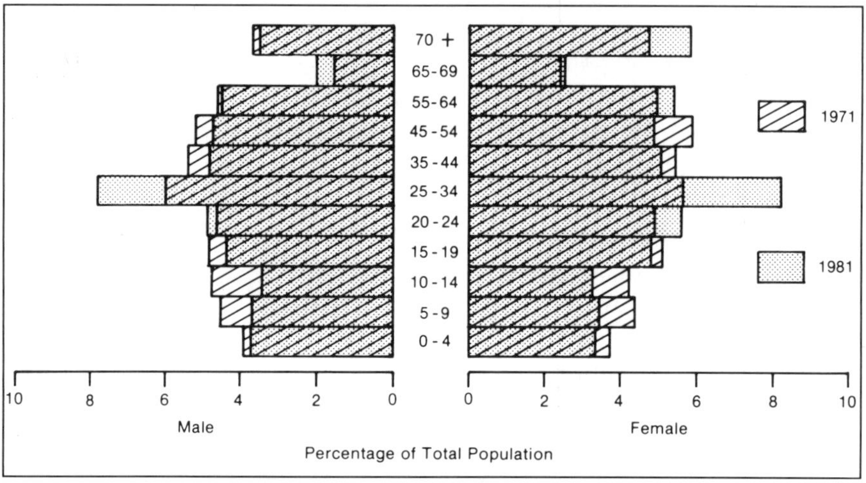


FIGURE 5:1
*Changes in Brandon's age structure: 1971-1981*

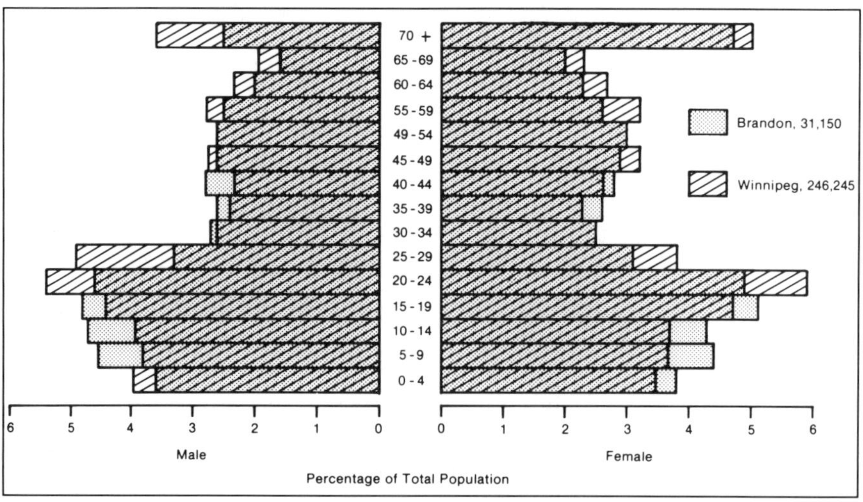


FIGURE 5:2
*A comparison of the age structures for Brandon and Winnipeg: 1971*

elsewhere. However, it seems likely that economic recession has reduced the availability of employment elsewhere, and that lower living expenses in Brandon have encouraged younger people to remain in the city. The other notable change in Brandon between 1971 and 1981 was the reduction in the number of children in all age categories. This is in keeping with the trend common throughout Canada of declining birth rates during this period. It is indicative of such factors as the economic recession and the increasing number of women in the labour force.

Residential growth in Brandon occurred predominantly to the south, east and west of the CBD, centred on Rosser Avenue (Chapter 4). Recent planning policies have directed residential growth away from the newly established industrial park in the eastern part of the city and have attempted to direct it westwards (Chapter 9). Thus, present residential development extends beyond Thirty-Fourth Street West (which represents the western boundary of the city in this study). Recent residential growth above the floodplain on the north side of the Assiniboine River has also occurred (Chapter 4).

Factorial ecology studies have identified relevant social variables which are appropriate indicators of social, economic and ethnic status, including indicators such as age characteristics, family structure, female characteristics (age and labour force participation of women), dwelling type and tenure, household and dwelling size, labour force participation, educational level, income, and ethnic characteristics. Thirty-one variables representative of these indicators were selected from the census data for use in the Brandon study (Hanson 1984). These were subjected to a principal components analysis, which on the basis of varimax-rotated factor loadings allowed identification of seven components each explaining over 5 percent of the total variance. The labelled components, which account for 80.9 percent of the total variance, are shown in Table 5:1. The component scores were then mapped to allow analysis of the spatial dimensions of the results. The five most important components will be discussed in turn and the results compared with findings from other, larger, western Canadian cities.

## Family Status Component

The principal component accounting for 25.4 percent of the total variance is labelled "family status." Variables which have the highest

TABLE 5:1
SUMMARY OF VARIMAX-ROTATED FACTOR LOADINGS

| | Positive Loadings | | Negative Loadings | Percentage of Total Variance |
|---|---|---|---|---|
| **Component I** (Family Status) | | | | 25.4 |
| Number of Children | 0.918 | Small Household | -0.794 | |
| Family Size | 0.910 | Old Age | -0.745 | |
| Percentage of Children | 0.887 | Middle Age | -0.733 | |
| Large Household | 0.872 | | | |
| Persons/Household | 0.859 | | | |
| Mature Adults | 0.675 | | | |
| Fertility Ratio | 0.591 | | | |
| Dwelling Size | 0.519 | | | |
| **Component II** (Household Type) | | | | 12.4 |
| Single Detached | 0.946 | Apartments | -0.746 | |
| Owner-Occupied | 0.934 | Percentage of Females | -0.621 | |
| **Component III** (Ethnic Status) | | | | 10.8 |
| Other Ethnic | 0.925 | British Origin | -0.929 | |
| Limited Education | 0.625 | | | |
| No Bath | 0.567 | | | |
| **Component IV** (Economic/Educational Status) | | | | 9.8 |
| University Graduates | 0.855 | Lodgers | -0.643 | |
| Average Income | 0.802 | Limited Education | -0.530 | |
| Dwelling Size | 0.558 | | | |
| **Component V** (Age/Marital Status) | | | | 9.1 |
| Percentage Single | 0.841 | Mature Adults | -0.598 | |
| Young Adults | 0.543 | | | |
| **Component VI** (Employment by Gender) | | | | 6.9 |
| Males Employed | 0.813 | Females in Labour Force | -0.695 | |
| **Component VII** (Housing Standards) | | | | 6.6 |
| High-School Graduate | 0.743 | | | |
| No Toilet | 0.733 | | | |
| No Bath | 0.667 | | | |

positive loadings on this component are associated with family characteristics, such as number of children, family size, large households and dwelling sizes. High negative loadings relate to variables such as small households (one- or two-person households) and proportion of individuals over 65 years of age. The emergence of "family status" as the principal component is consistent with other western Canadian studies (Peucker and Rase 1971; Davies 1975, 1978).

The spatial variance in family status in Brandon is shown in Figure 5:3. The component scores differentiate between areas with well-defined family indicators, which are identified as areas with high scores on this component, and those areas where family structure is poorly developed, which received low scores. Medium scores reflect average or transitional areas. In other cities, family status has exhibited a generally concentric pattern, with the city centre containing a population with least well developed family characteristics and the edge having the most developed family structure. In Brandon these trends are generally evident, although the pattern is distorted in some areas because of certain place-specific features. Lowest scores for family status occur in two separate core areas. As in other cities, few families live in the CBD, the main function of the area being commercial. Housing within the

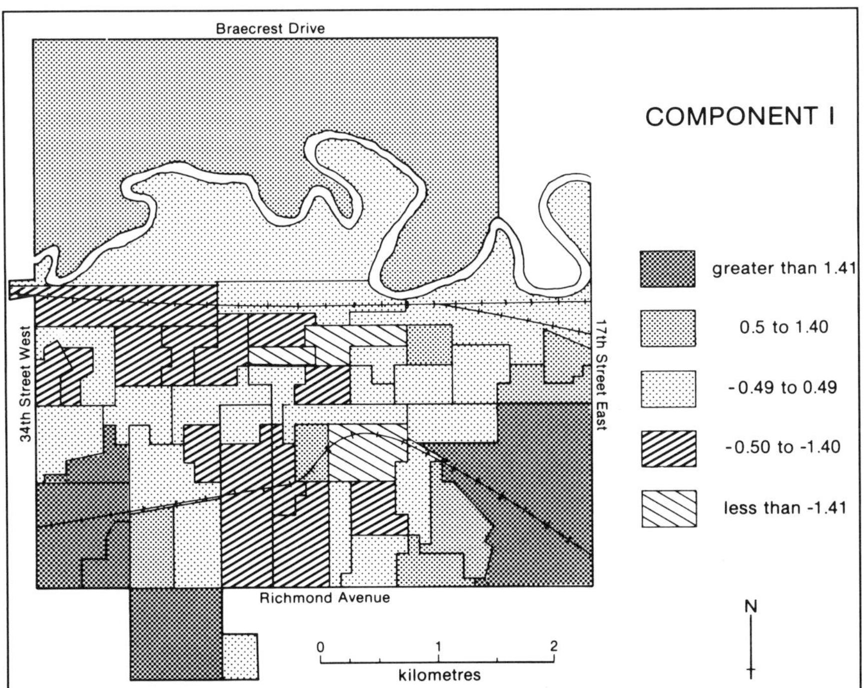


FIGURE 5:3
*Spatial variance in "family status" in Brandon*

CBD is either in the form of apartments above commercial establishments or large older homes that have been converted into apartment units. Neither of these two housing types is popular as a family environment. The second area of low family status is located south of the core area around the CNR yards and warehouses. A small cluster of houses is situated in this area consisting of a diverse assortment of apartment units, two-family dwellings and single-family dwellings. The low family status scores in this area may be attributed to the presence of the rail yards and the warehouses, neither of which adds to its attractiveness as a family setting.

Adjacent to each area of low family status, two axes of slightly higher status extend south and west from the core areas to the edge of the city. The axis to the south is easily explained as there is very little residential development in this area which contains the Provincial Exhibition Grounds and the Agricultural Extension Centre. The presence of Fairview Home, a nursing home for the elderly and for the disabled, also contributes to the lower family status score. The westward trending axis consists of one of the older residential areas in Brandon (Chapter 4). Most of the households are well established with a stable population, which is reflected by the large numbers of middle and old age adults, and the lack of children. It is likely that this area contains a larger proportion of households in which the children have already grown and left home. Brandon University is also located in the heart area. The presence of the university and the high number of single students, many of whom live in campus residences, in apartments, or in lodgings in larger houses also detracts from the family status component. The two areas of fairly low family status extend from near the city centre to the outer limits of the city. This pattern is not consistent with the findings of studies in other cities, where it is most common to find an overall increase in family status levels as one moves outwards from the centre of the city, creating a concentric pattern. There is no obvious reason why Brandon should differ in this respect, although it may be a function of the city's smaller size.

The areas of high family status occur, as expected, on the suburban fringes of the city, the two highest areas of family status being in the southwestern and southeastern corners. Housing in the southwestern part of the city consists of recently constructed single-family homes designed for family living. The southeastern area includes an industrial park, and nearby housing has locational appeal for employees. The CNR line and a strip of open land provide an efficient buffer between industrial and residential areas. Although both areas reflect similar "family status," the areas appear distinct in terms of "social status," with the southwest area tending to house middle-class residents and the southeast a working-class population.

Another area in which family status scores are relatively high is the Kirkaldy Heights subdivision (Figure 4:4), although scores are not as high as one might expect. Located north of the Assiniboine River, this area has several features which make it attractive to families, such as large tracts of open unused parkland. However, the development plan for the area proposed in 1968 designated a mixed housing stock of duplex housing and multiple apartment units which attracts single residents as well as families.

Two apparently anomalous small pockets of relatively high family status are also evident adjacent to the core areas of low family status near the city centre. Like other urban centres, the preferred housing areas when Brandon was first developed were close to the downtown area (Chapter 4). Many of these old houses are still sound in construction, and appealing both in construction and design. Although some of the larger houses have been converted into multi-unit dwellings, many still remain as single-family dwellings, some of them having remained in the same family for several generations.

As in studies of other western Canadian cities, differences in family status explain the greatest amount of variance between areas within Brandon. The concentric pattern displayed in other studies, with an increase in family status away from the city centre, is also evident to some degree in Brandon, although site-specific variations reflect Brandon's individual character. These differences in family status are evident in the physical appearance of neighbourhoods, as single-family houses are the preferred housing type for family units.

## Household Type Component

In studies in other cities, variables associated with household type usually have been loaded on the same component as family status variables. However, in Brandon, variables associated with household type load on a separate component which accounts for the second highest proportion of the explained total variance. The "household type" component accounts for 12.4 percent of the total variance and is characterized by high positive loadings of variables relating to the presence of "single detached" and "owner-occupied" homes, and high negative loadings on the variable, "number of apartments." The "percentage of females" variable also has a high negative loading on this component although there is no obvious explanation for this relationship with "household type" variables.

Although the "household type" component is extracted as a component separate from the "family status" component, similar spatial patterns are shown (Figure 5:4). The central area of Brandon is characterized by low scores on the "household type" component, reflecting the higher number of apartments in this area, and there is generally an increase in the scores as one moves away from the centre reflecting increasing numbers of single detached and owner-occupied homes. Pockets of low scores on the "household type" component are found in some peripheral areas reflecting the

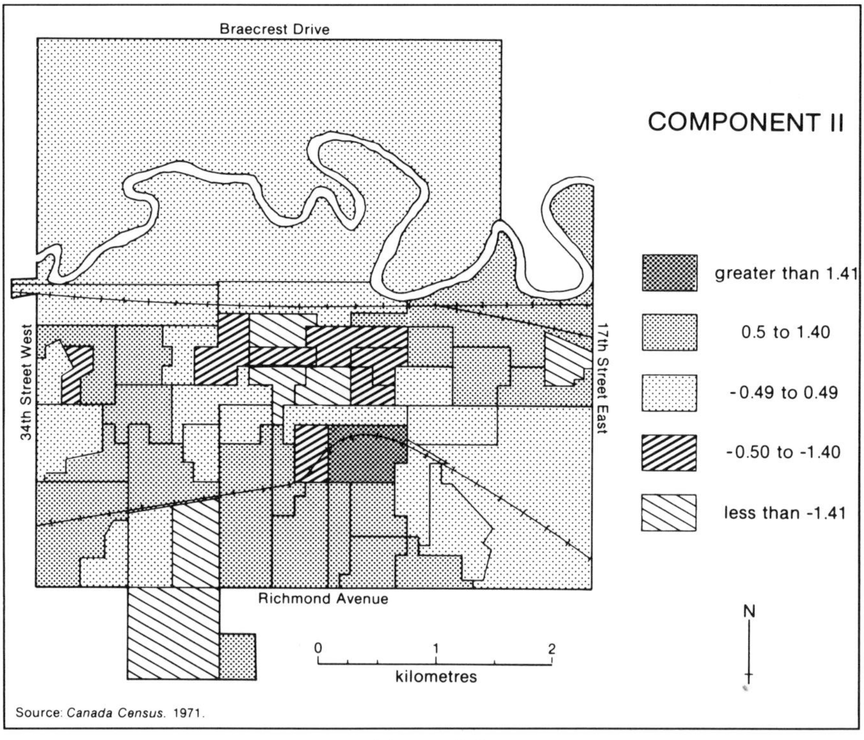


FIGURE 5:4
*Spatial variance in "household type" in Brandon*

presence of apartments. These suburban apartments reflect the city's zoning of the areas as multi-family districts. Similar zoning policies occur in most North American cities, but the effects are probably more apparent in Brandon because of the smaller size of the city where the policies probably have a greater impact on community patterns and development (Chapter 9).

## Ethnic Status Component

The third component, which explains 10.8 percent of the total variance in the data set, is labelled "ethnic status." The variable with the highest loading on this component is "other ethnic," which refers to any ethnic origin other than British or French. This variable is inversely related to the other strongly loading variable, "percentage of British origin." The spatial patterns of ethnic status in Brandon are shown in Figure 5:5, with high scores identifying those areas with high proportions of residents who are not of British

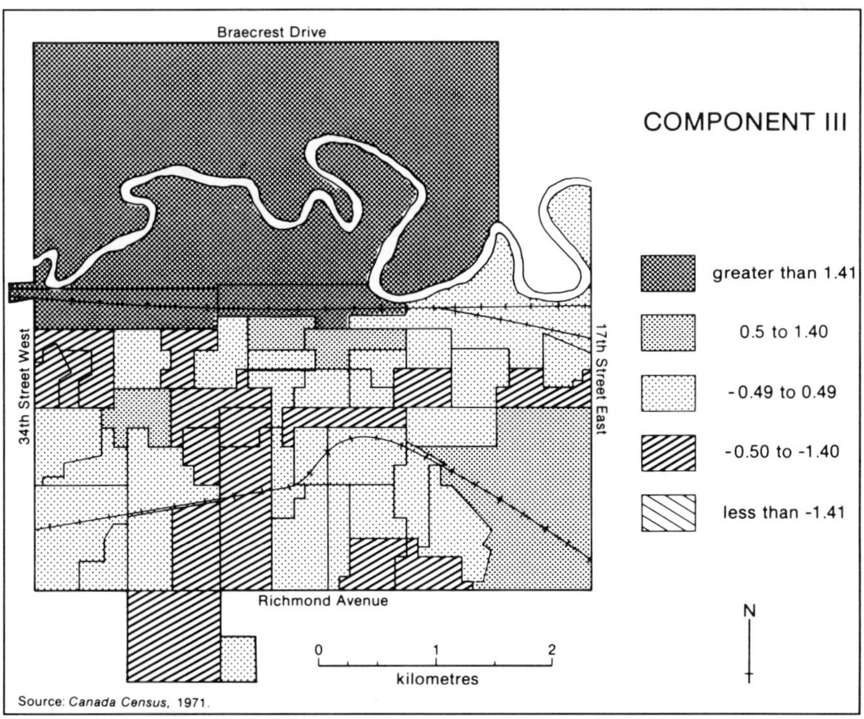


FIGURE 5:5
*Spatial variance in "ethnic status" in Brandon*

origin. Brandon, in common with other western Canadian cities, displays a multi-nucleated pattern of dispersion.

In Brandon, the dominant area of ethnic concentration is north of the CPR line, an area which has traditionally been characterized by people of East European descent (Clark and Stadel 1972). In the early years of settlement, East European immigrants were of lower economic status and were forced, through various social pressures, to occupy the less desirable land located on the floodplain (Hume 1981). Although the social climate of Brandon has changed over the years, the ethnic distribution and cultural landscape of this area still reflect the historical patterns of residence. Churches, social clubs, and house and garden styles attest to a continuing sense of East

Plate 5:1

*The Assiniboine Gardens area of Brandon, located between the CPR tracks and the Assiniboine River, is the only true ethnic area of Brandon. In the right foreground is St. Mary's Ukrainian Catholic Church with the Ukrainian Reading Association to the left of it; in the middle is the Ukrainian Greek Orthodox Church of Holy Ghost with the Polish Sokol Association's Club and Hall to the south of it. The unwooded area in the middle ground marks the former course of the Snye River which used to enter the Assiniboine off the photograph to the right. Close to the Assiniboine are the Manitoba Pool Livestock Market and the Burns Meat Plant. The North Hill is seen in the background. (John Welsted, 1987)*

European cultural identity (Plate 5:1). Other areas north of the Assiniboine River floodplain, including the Kirkaldy Heights district, also have high scores on the ethnic status component. This may indicate that even beyond traditional areas of settlement, the desire for cultural commonality still affects residential choices.

There are two other areas of ethnic concentration in Brandon, one in the southeast part of the city, and the other in the west. Each of these areas is presently, or was at one time, located adjacent to a rail line, indicating that they may have been deemed less desirable as residential areas by other social groups in Brandon. However, for the East European immigrants, the railway often provided a source of employment; thus, locating near the rail line may have been construed as a benefit rather than as a social constraint. Again, these areas of ethnic concentration have remained resilient over time.

## Economic/Educational Status Component

The fourth component, which is labelled "economic/educational status," accounts for 9.8 percent of the total variance. Highest positive loadings on this component include "percentage of university graduates," "average income," and "dwelling size"; whereas negative loadings include "educational levels below Grade 9," and the "percentage of households with lodgers." The spatial configuration of this component (Figure 5:6) exhibits a clear distinction between the east and the west sides of the city. The east end is an area of low economic/educational status, while the west exhibits scores that are significantly higher. This distinction reflects city planning decisions which have allocated major industrial development to the east side of the city while residential growth is occurring in the west and southwest (Chapter 4). This pattern of land use, common to many North American cities, is in part an environmental response to prevailing westerly winds (Chapter 3). The west end, because of its recent development and high proportion of newly constructed dwellings, is attractive to individuals who can afford such housing. Consequently, a high proportion of residents in this area is well educated and engaged in managerial and professional occupations. In contrast, the east end consists of many older and less expensive homes which are more attractive to blue-collar workers. Further, many are employed in the Green Acres Industrial Park complex (Figure 4:5); thus proximity to the workplace is an added attraction. The northern section of Brandon is also an area of relatively high

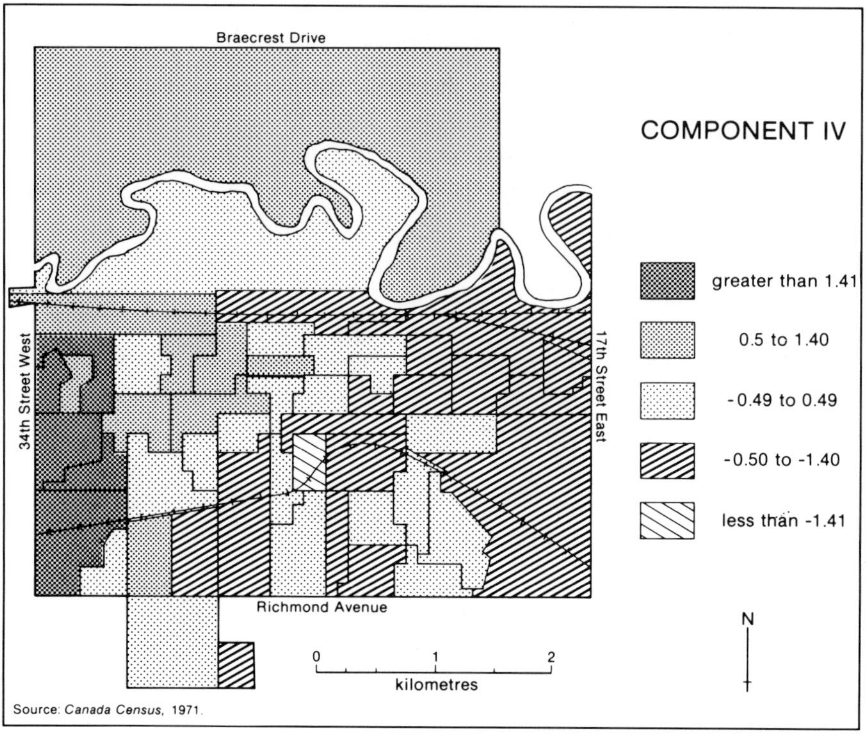


FIGURE 5:6
*Spatial variance in "economic/educational status" in Brandon*

economic/educational status which, as in the west end, can be attributed to the construction of relatively new and expensive housing. Other studies reveal that economic status normally displays sectoral patterns. In Brandon this is true in a general manner, although because of the relatively small size of the city the sectors are limited in number.

## Age/Marital Status Component

The fifth component accounts for 9.1 percent of the total variance of the data. Based on the high positive loadings on this component of variables such as "percentage of single residents" and "percentage of young adults" and the high negative loadings on "mature adults," the component is labelled "age/marital status." This component exhibits a diverse pattern, with small pockets of both high and low scores scattered throughout the city (Figure 5:7). A notable area of young single adults is the central area of the city. Apartment units

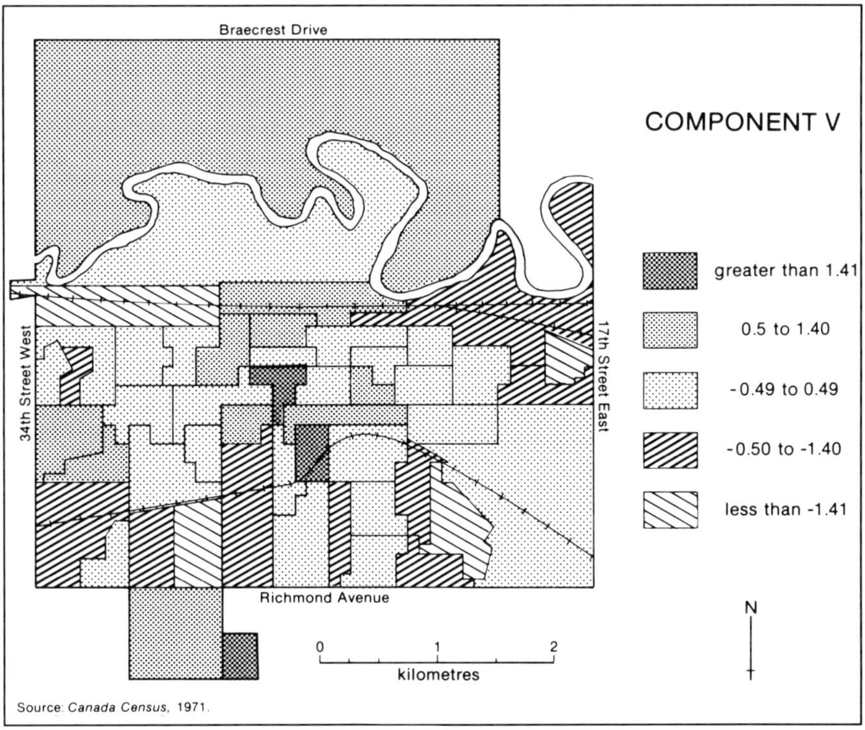


FIGURE 5:7
*Spatial variance in "age/marital status" in Brandon*

in this area meet the housing needs of single residents, who for social reasons are less attracted to suburban locations. This area also serves the needs of university students because it is readily accessible to Brandon University. An area in the extreme southern portion of the city also has a high proportion of single adults. There are two large mobile home parks in this district, and it is possible that this type of low-cost, temporary housing appeals to young single adults who generally have lower incomes. Other patterns displayed by this component are difficult to interpret because of their dispersed nature but this component is interrelated with other components, especially those associated with family status and household types.

## Discussion

The five main components extracted for Brandon, which account for over 60 percent of the overall explanation in the variance of the data, are similar to those found in other Canadian studies. Four of the five components are almost identical in character to those noted in previous studies (Davies and Barrow 1973; Davies 1975, 1978; Murdie 1969; Nicholson and Yeates 1968; Peucker and Rase 1971). These components are "family status," "economic status," "household status" and "ethnic status." The two components listed in Table 5:1, and which are not discussed here, "employment by gender" and "housing standards," are not as clearly defined. The general similarity in the definition of Brandon's social areas to those in other cities substantiates the theory that there is a common underlying dimension of community structure in North American cities which is a composite of the four major components. As noted earlier, this structure exhibits a concentric form for family and household status, sectoral patterns for economic status and multi-nucleated patterns for ethnic status. These patterns are exhibited in a general way in Brandon, although as expected there are distortions from the idealized model. Discrepancies, which are also characteristic of studies elsewhere (Davies 1975, 1978), reflect the unique features of Brandon's setting and evolution, such as the presence of the Assiniboine River and floodplain, the location of railway lines, and historical/political development decisions (Chapter 4).

**Acknowledgement**

Most of the information included in this chapter was collected by Randy Hanson for an undergraduate thesis, part of the requirements for a four-year Specialist degree in Geography at Brandon University.

**References**

Clark, L. and C. Stadel. 1972. Brandon. In *Southern Prairies Field Excursion Tour Guide.* Regina: University of Saskatchewan, Regina Campus.

Davies, W.K. 1975. A multivariate description of Calgary's community areas. In B.M. Barr, ed., *Calgary Metropolitan Structure and Influence.* Victoria: University of Victoria, Western Geographical Series, Volume 11.

Davies, W.K. 1978. A social taxonomy of Edmonton's community areas in 1971. In P.J. Smith, ed., *Edmonton the Emerging*

*Metropolitan Pattern.* Victoria: University of Victoria, Western Geographical Series, Volume 15.

Davies, W.K. and G.T. Barrow. 1973. A comparative factorial ecology of three Canadian prairie cities. *The Canadian Geographer,* 17(4):327-51.

Dominion Bureau of Statistics. 1971. *Enumeration Area Printout, Brandon.* Ottawa: Queen's Printer.

Hanson, R. 1984. Brandon's Community Areas: A Factor Analytical Study. B.A. thesis, Department of Geography, Brandon University.

Hume, M. 1981. *Brandon a Prospect of a City.* Brandon: City of Brandon.

Murdie, R.A. 1969. *Factorial Ecology of Metropolitan Toronto, 1951-1961.* Chicago: University of Chicago Press, Department of Geography Research Paper, Number 116.

Nicholson, T.G. and M.H. Yeates. 1969. The ecological and spatial structure of the socio-economic characteristics of Winnipeg, 1961. *The Canadian Review of Sociology and Anthropology* 6(3):162-78.

Peucker, T.K. and W. Rase. 1971. A factorial ecology of Greater Vancouver. *Contemporary Geography: Western Viewpoints,* B.C. Geographical Series 12:81-96.

# SIX
# Brandon's Manufacturing Sector

*Paul Curtis*

Founded in May 1881, the city of Brandon was incorporated the following year (Chapter 4). A contemporary wrote:

> The Brandon to which I came in March 1883 was a different Brandon from which it had been in April 1882. It didn't look finished; but it carried a less transitory air. . . . At the base of everything was a great country for which Brandon was the market centre. . . . Being a market place for all supplies for so much country, Brandon was the Mecca for streams of horse and horned traffic; most likely larger than anything that has come to any place in Canada since railways made it unnecessary for produce to be teamed into Toronto from around Lake Simcoe and Georgian Bay. Neither Winnipeg nor Portage la Prairie ever had so much farming country directly tributary to it as Brandon did until branch lines began to remove the Pacific Avenue elevators, which, for several years were our outstanding architecture. (Trotter 1925)

How much of this is fact and how much fancy it is not possible to tell. However, Trotter's observations tend to support Stanford Fleming's prediction that the location was promising for the growth of a city:

> I have carefully examined all data at command. . . . If the line be carried to a point in the valley of the Assiniboine, near the mouth of the Little Saskatchewan, where the land remains unsurveyed and ungranted, there might here be established the site for a city, which would shortly become important . . . and from which desirable colonization lines might, in the near future diverge (1) to the north-west, (2) to the west, and (3) to the south-west; and thus the projected city would become an important railway and business centre. (Fleming 1879)

In the passages quoted, one fact is immediately apparent: the birth and growth of the city was as much a result of its hinterland as it was of any inherent site advantage; site was subservient to situation. The importance of situation has not changed during the city's history as evidenced by the opening paragraphs of a 1985 publication by Brandon's Industrial Commission which extol the transport links and the prosperous agricultural hinterland, and which make the unequivocable statement that "Brandon is the regional centre of Western Manitoba."[1] The factors of site and situation are paramount in any study of Brandon's commerce and industry. As the situation changes, the commercial and manufacturing sectors respond accordingly; for example, Trotter reported the loss of grain elevators from the city as branch lines were constructed during the 1880s. A century later, the wheel turned full circle. Abandonment of branch lines within the Brandon catchment area led to the opening in 1983 of an inland terminal type of grain elevator, not on the original Pacific Avenue site but nevertheless within the city.

A 1913 city publication extolling the virtue of Brandon as a manufacturing site observes that "manufacturers locating in Brandon will have an immense advantage in freight rates in all lines in which heavy materials must be brought from the east."[2] Also, because Brandon was made a divisional point of the Canadian Pacific and Great Northern Railways, and the Grand Trunk Pacific completed a line to the city connecting American and Canadian railways, Brandon became the district centre for some 32,500 square miles (84,175 square kilometres) wherein were located some 289 small towns and 3,500 retail merchants. The importance of Brandon's regional setting is apparent.

An interesting comment from the same document was a somewhat ambivalent request for capital investment:

> There has never been a large amount of capital invested in any of the Brandon industries. The great success that some of them have achieved without capital is an indication of what might be accomplished by manufacturers coming to Brandon with ample capital.[3]

Here, surely, is free enterprise at its finest. The statement makes no reference to grants, tax forgiveness, serviced lots, or any other form of inducement to locate within the city. Even so a substantial number of products were manufactured or assembled in Brandon at the time, including gasoline engines, windmills, pumps, fire engines, portable sewing machines, steel plate hot furnaces, trip hammers, store counters, show cases, school desks, church pews, pulpits, mattresses and lightning rods. Also important were flour grinding and the making of oatmeal.

Flour and oatmeal are noteworthy in that Brandon, from its earliest days, became known as the "Wheat City" (Chapter 4) indicating its dependence on the agricultural hinterland. Manufacturing within the city has always reflected this dependence, as indicated by a 1984 article, first published in *Trade and Commerce*,[4] which lists agricultural grains, special crops including oil seeds, potatoes and livestock as four of the raw materials available for processing and further manufacturing, this in a total list of twelve items. Regardless of the year of publication, information distributed by the city advertising its assets always includes the acreages under various crops, the total number of various animals, and other "agricultural" information about the surrounding area. Today a livestock yard, a meat processing plant, a waterfowl processing plant, seed and feed mills are all direct responses to the availability of agricultural raw materials. Further confirmation of Brandon's reliance on these raw materials for its industries came from the city's industrial commissioner in 1985, when he stated that "twenty out of our 29 fabricating and processing plants are directly related to the farm community."[5]

Classical industrial location theory (Smith 1971) identifies six major factors which manufacturers consider in any potential site: availability of raw materials, transportation, labour, power, availability of capital, and markets. Assessing these factors results in the conclusion that Brandon possesses no inherent advantages for large-scale manufacturing, a conclusion that is confirmed when the historical record of the number of manufacturing plants is examined. In 1923, the first year for which records are available, the city contained thirty-five plants employing 484 persons. By 1930, the number of plants had risen to thirty-six but the number of employees had decreased to 433 (Figure 6:1). The number of manufacturing plants has been remarkably stable; three times, in 1937, 1941 and 1946, the number of plants fell to twenty-nine and did not exceed fifty until 1980. The long-term average number of manufacturing plants in the city is thirty-six with an average of

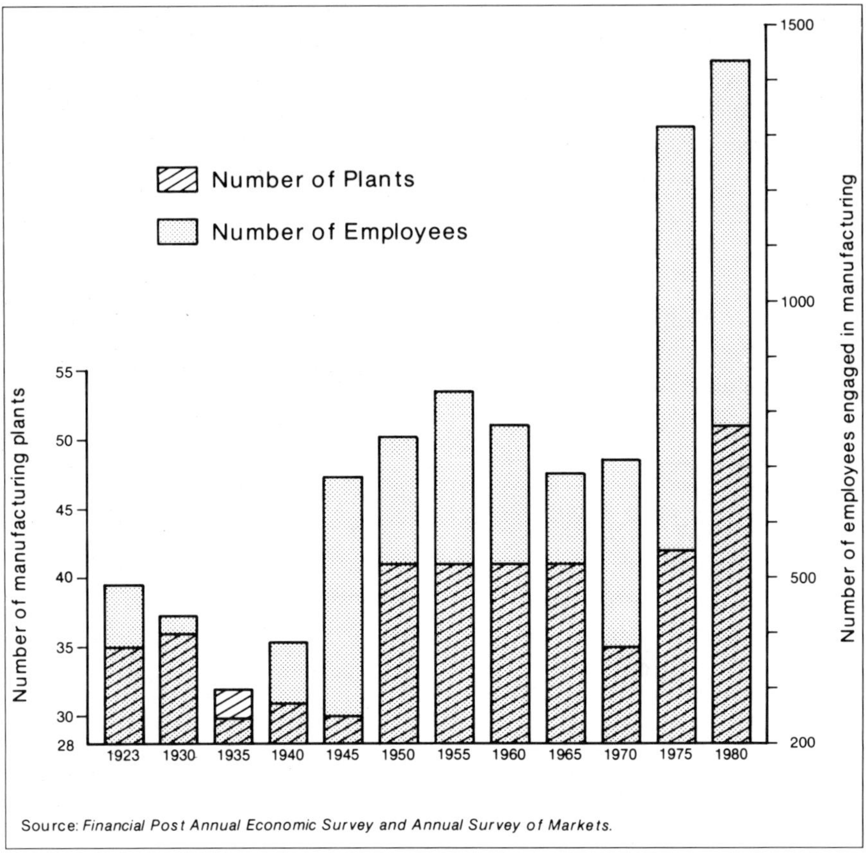


FIGURE 6:1
*Number of manufacturing plants and number of employees engaged in manufacturing in Brandon: 1923-1980*

thirteen employees per plant. While averages can be misleading, they give an accurate picture in this case; that is, of few plants with few employees in each. A recently published list of Brandon industries shows twenty-five of the forty-eight firms listed having between twenty and twenty-five employees.[6]

When the percentage of the total labour force engaged in manufacturing is compared with that in Canada as a whole, two observations can be made (Table 6:1). First is the relative stability of the Brandon manufacturing work force and second, that Brandon has a lower percentage of its work force engaged in manufacturing.

Table 6:1
Percentage of Brandon's Labour Force Engaged in Manufacturing Compared to that in Canada

| | 1931 | 1941 | 1951 | 1961 | 1971 | 1981 |
|---|---|---|---|---|---|---|
| Brandon | 10.5 | 11.2 | 10.6 | 11.1 | 9.0 | 10.0 |
| Canada | 16.0 | 21.4 | 15.8 | 21.7 | 19.6 | 19.1 |

*Source:* Dominion Bureau of Statistics (DBS) 1931-1981.[7]

Percentage change in the manufacturing labour force can also be compared with Canada as a whole (Table 6:2). Variations within these figures reflect business cycles and also the rate of change of the population as a whole.

Table 6:2
Percentage Changes in Brandon's Labour Force Engaged in Manufacturing Compared to those in Canada

| | 1931-41 | 1941-51 | 1951-61 | 1961-71 | 1971-81 |
|---|---|---|---|---|---|
| Brandon | 19.6 | 13.5 | 36.32 | 9.1 | 46.1 |
| Canada | 53.5 | -13.2 | 66.9 | 13.4 | 43.0 |

*Source:* DBS 1931-1981.

A further indicator of the place of manufacturing is the percentage change of numbers employed in all industries compared with the national situation (Table 6:3).

Table 6:3
Percentage Changes in Brandon's Total Labour Force Compared to those in Canada

| | 1931-41 | 1941-51 | 1951-61 | 1961-71 | 1971-81 |
|---|---|---|---|---|---|
| Brandon | 12.2 | 19.8 | 31.1 | 34.7 | 30.9 |
| Canada | 14.8 | 17.3 | 22.2 | 25.4 | 46.3 |

*Source:* DBS 1931-1981.

Of note in Table 6:3 is the fact that for three of the five-year periods for which figures are available, Brandon's total labour force increased, on a percentage basis, more than the national force. However, a comparison of Table 6:2 with Table 6:3 shows that in only two of the five decades did employment in manufacturing within the city increase at a greater rate than in the nation as a whole. Such comparisons lead to the conclusion that Brandon's labour force is primarily employed in tertiary and quaternary activities rather than in manufacturing, confirmation of which is the observation that over the long term the two biggest employers in the city have been its hospitals and its public schools.

Of particular significance to the welfare of any community is the number of basic and nonbasic activities. Nonbasic activities are those which are necessary for the community to be regarded as self-sufficient; that is, they serve the local market. Where a community is not self-sufficient in a commodity, it must import that commodity resulting in an outward flow of capital. Conversely, basic industries are capital earners; that is, the community produces more of the commodity than it requires and exports the surplus. The location quotient,[8] derived from a simple ratio comparison, attempts to separate basic and nonbasic industries, a quotient of less than one indicating a deficit of production while a quotient greater than one indicates a surplus. Table 6:4 is a summary of Brandon's basic industries for a period of sixty years.

The table shows the long-term value of agriculture to the city's manufacturing scene. Despite changes in categories over the years, the constancy of the food and beverage industries is noteworthy as is the category which included iron and its products in 1931 and metal fabrication in 1981. Whereas Franklin stoves, trip hammers and so on were important manufactured items in the city in 1913, farm truck boxes, steel grain bins and steel machine sheds took their place in 1981. At both times, the market for the metal products was local rather than national, national rather than international.

Interesting anomalies in the figures can be explained by tracing the rise and fall of specific industries. The decline of the "petroleum and coal" category between 1961 and 1971 can be explained by a change of category; that is, what was included within the industrial classification, and more importantly, by the demise of the Brandon Petroleum Refinery during this decade and the associated loss of employment (Plate 6:1). On the other hand the appearance of "chemicals and allied industries" as an important local basic industry during the same decade reflects the arrival of the Simplot Chemical Fertilizer Plant and of Hooker Chemicals.

## Factors of Industrial Location

An examination of the six major factors of industrial location reveals why Brandon has difficulty in attracting industry.

TABLE 6:4
BRANDON'S BASIC INDUSTRIES

| | Industrial Subgroup | Location Quotient |
|---|---|---|
| 1931 | Vegetable Products | 1.525 |
| | Animal Products | 1.612 |
| | Iron and its Products | 1.743 |
| | Non-metallic Minerals | 1.436 |
| 1941 | Vegetable Products | 1.756 |
| | Animal Products | 2.292 |
| | Iron and its Products | 1.133 |
| | Non-metallic Minerals | 1.900 |
| 1951 | Food Products | 2.413 |
| | Liquor and Beverages | 3.500 |
| | Fur and Fur Products | 4.037 |
| | Printing, Publishing & Bookbinding | 1.235 |
| 1961 | Food and Beverages | 2.589 |
| | Printing, Publishing & Allied | 1.768 |
| | Metal Fabrication | 1.648 |
| | Petroleum and Coal | 11.570 |
| 1971 | Food and Beverages | 2.238 |
| | Printing, Publishing & Allied | 1.719 |
| | Metal Fabrication | 1.272 |
| | Non-metallic Minerals | 3.924 |
| | Petroleum and Coal | 2.034 |
| | Chemicals and Allied | 3.797 |
| | Fertilizer | 19.700 |
| 1981 | Food and Beverages | 1.551 |
| | Tobacco | 2.220 |
| | Textile | 1.135 |
| | Printing and Publishing | 1.753 |
| | Metal Fabrication | 1.840 |
| | Electrical Products | 1.570 |
| | Chemical and Allied | 3.390 |

*Source:* Derived from DBS, 1931-1981.

### *Raw Materials*

With the exception of agricultural products, the raw materials readily available in the Brandon area listed by *Trade and Commerce*[9] are virtually ubiquitous for Canadian industry; salt, clay, sand and gravel are not normally major drawing cards for manufacturing plants.

PLATE 6:1

*The junction of Eighteenth Street and Richmond Avenue in 1960. The photograph, taken looking southeast, shows the Brandon Petroleum Refinery on the southeast corner. The refinery closed during the 1960s and has been replaced by a garage, a gas station, a fast food outlet and a Canadian Tire store. The southwest corner, which was being used for agriculture when the photograph was taken, is now the location of the Brandon Shoppers' Mall. (Public Archives of Manitoba: Brandon — Business — Anglo-Canadian Oil 5)*

While the linkage between the meat-packing plant and the agricultural base is obvious, the linkage between the agricultural raw material base and the Ayerst Organics plant is less clear. This firm, a subsidiary of an international pharmaceutical company, extracts estrogens from pregnant mares' urine. Before locating in the city in 1966, it undertook an extensive survey of Brandon's rural hinterland to assess the viability of the raw material source. Ultimately an agriculturally based industry, it illustrates the efforts that have been made to broaden the agricultural raw material base of Brandon's industry. It is an attempt to smooth out fluctuations in the supply of agricultural raw materials which inevitably occur when the supply is at the mercy of the elements.

Since the city and its environs have few raw materials of their own, manufacturers who consider Brandon as a potential site are faced with importing raw materials, so adding to the cost of production.

*Transportation*

In large measure Brandon owes its existence to its role as a transportation node. It was the CPR making Brandon a divisional point which gave the city an initial advantage over its nearest competitors for urban supremacy in the region (Chapter 4). The railroads continue to play an important role within the city's economy. Brandon remains a divisional point for the CPR and within the city is a thirty-track marshalling yard, the terminal for 450 miles (724 kilometres) of main line track and more than 1,000 miles (1,609 kilometres) of secondary track. The CNR also has a small marshalling yard and provides industrial sidings to customers. Examples of local industries benefitting from the railroad sidings are Simplot Chemicals, Cargill Nutrena Feeds Ltd., and Burns Meats Ltd. Both railway organizations provide daily service to east and west, but both systems rely much more heavily on long distance hauls as a major source of traffic. As compensation for reduced local rail service in recent years, each system provides daily trucking services as well as "piggy back" services to the city and its hinterland.

Throughout its history, Brandon has never been perceived by potential manufacturers as a break-in-bulk point. This role has been captured by Winnipeg in the east and Regina in the west. Winnipeg, only 208 kilometres distant, developed primarily as the gateway to the Prairies and is an almost insurmountable intervening opportunity for manufacturers whose potential market area is the eastern Prairies. Regina, 365 kilometres away, is a more attractive location for manufacturers interested in serving the central prairie region. South of Brandon, there are no major centres on either side of the Canada/United States border (only 106 kilometres distant) and the major highway, Highway 10, joining the two countries is normally devoid of heavy trucks. To the north, the picture is similar (Chapter 10), with Dauphin (population in 1981, 8,971; distance, 167 kilometres) being the closest centre of any size. Interaction between the two centres with respect to manufacturing is negligible. Brandon's problem is illustrated by a statement in the city's 1985 publication, *Brandon, Manitoba, The Place to do Business*, which refers to Brandon at the centre and states that: "Major continental traffic routes pass through the city, the Trans-Canada Highway, both Canadian Trans-Continental railways, and No. 10 Highway north-south which extends into the United States as Highway No. 281." Unfortunately for the city, the operative words in the statement are "pass through."

Brandon's poor location as a potential break-in-bulk point is illustrated by its inability to attract and maintain air traffic. Despite the best efforts of several city administrations, it was not until 1981 that the city was able to attract a national jet service. This two-way daily service between Toronto and Vancouver calling at Brandon immeasurably increased national accessibility to the city. However, in the spring of 1988 the airline company, citing poor economic returns, announced plans to downgrade its service from a national to a regional operation whereby Brandon commuters would be transferred to Winnipeg for national flight connections.

The inability of the city to become part of the national air network can only have detrimental consequences for its manufacturing base. Since manufacturing plants attracted to Brandon are likely, for the forseeable future, to be either footloose industries or branch plants of larger national or international companies, rapid access to both parts and personnel is essential. The continuous efforts of the city administration and the Brandon Chamber of Commerce over the years to attract a national air carrier vividly attest to their realization of the need for direct accessibility to the Canadian manufacturing scene.

Also, several attempts have been made to secure regional air service between Brandon and Winnipeg. All of them have failed. The reason most often given by Brandon business people is that unless a direct transfer flight from Winnipeg is the objective of the journey, the inconvenience of getting to and from the respective airports far outweighs the convenience of letting someone else, the pilot, do the driving, particularly when there is no time advantage involved.

While further exacerbating the problems of regional airlines, the twinning of the Trans-Canada Highway between Brandon and Winnipeg cut approximately one-half hour from travel time between the two centres. The importance of this transport link was evidenced by a 1969 transportation study of the Westman Region of which Brandon is the centre.[10] Of the more than one million tons of freight accounted for by this study, approximately 88.4 percent originated in Winnipeg[11] and of the 130,000 tons originating in the Westman study area almost half were destined for the provincial capital. The absence of manufacturing raw materials transported is obvious even from a cursory look at the commodities handled. Building materials and machinery dominate inward flows while agricultural products dominate outward flows. Road service is provided to the city by twenty-five trucking services, five of which are locally operated.

The importance of the Assiniboine River as a means of transport during the early development of the city has been described elsewhere in this volume (Chapter 2). It is now of no significance for transportation.

According to Brandon's former industrial commissioner,[12] freight rates are neutral with respect to the city. While in the main this statement is undoubtedly true, some companies are provided with transport incentives by the trucking companies in order to reduce costly empty back hauls; this is particularly so in the case of international trucking.

Since Brandon has few raw materials other than those produced from the agricultural base, the transportation factor is probably the most important factor assessed by potential industries.

*Labour*

As an educational centre within Manitoba, Brandon ranks second behind metropolitan Winnipeg. Each year the University and Regional Community College graduate hundreds of well-trained students in a variety of trade and academic disciplines. The Community College is particularly responsive to perceived local community needs. The city also has a Regional Vocational High School which offers regular academic classes and initial training in a variety of practical trade skills. At the Brandon Mental Hospital a psychiatric nursing programme is offered while at the General Hospital a registered nurses programme graduates a steady stream of highly qualified nurses.

With all of the training institutions outlined, as well as others in the beautician and cosmetology fields, a large, varied and highly trained workforce is available. Surprisingly, in a survey of local establishments in 1978,[13] a poorly trained workforce and/or lack of a workforce was ranked first in a list of sixteen "problems encountered." Conversely, in a list of six named advantages, labour does not appear at all.

*Power*

Manitoba Hydro maintains a coal-fired steam generating plant within city limits (Chapter 2), while both the Trans-Canada Oil Pipeline and the Trans-Canada Natural Gas Pipeline pass a few kilometres north of the city. Each pipeline is tapped for city requirements.

The availability of natural gas in large quantities was a factor which the Simplot Chemical Company considered in choosing Brandon as a location. In the process of producing chemical fertilizer, considerable quantities of natural gas are used and by scheduling its peak manufacturing season during the city's low residential use season, the company was able to negotiate reduced rates with the gas company. The arrangement is also advantageous to the gas company because it allows for efficient use of its installed capacity.

*Availability of Capital*

According to the city's former industrial commissioner,[14] capital has not been much of an obstacle to industrial location within the city. Nevertheless, use of governmental and other sources of capital has had considerable influence upon the development of the city's industrial base.

An example is the Simplot Chemical Company which established in Brandon in 1965, underwent an expansion in 1983, and is the most capital-intensive company in Brandon (Plate 6:2). Capital incentives offered to encourage it to locate within the environs of the city included: (1) an outright grant of $5 million from the federal government for locating in a designated area; (2) a guaranteed loan of $20 million at low interest rates of 7 percent when comparable rates for industrial projects were 8 to 8.5 percent; (3) a further supplemental trust indenture allowing interim financing of up to $3.5 million from the federal government; and (4) the purchase of 1,100 acres of land from the Rural Municipality of Cornwallis for the nominal sum of one dollar per acre, the location of the land being immediately adjacent to the city's already established industrial park.

PLATE 6:2
*Part of southeastern Brandon is occupied by the Green Acres Industrial Park (in the middleground) and by the Simplot chemical plant (in the left foreground). The view is to the west and shows Victoria Avenue (on the right) and Richmond Avenue (on the left), east-west trending roads which are one mile apart. Assiniboine Community College is located at the junction of Victoria Avenue and Seventeenth Street East in the right middleground. (John Everitt, 1987)*

Approximately 10 percent of *Brandon, Manitoba, The Place to do Business* is devoted to information concerning financial incentives for which the city is eligible. However, all of these are either provincial or federal and are available throughout the province and are not, therefore, major incentives to locate specifically in Brandon. Provincial legislation forbids the manipulation of local tax legislation in order to attract industry.

Although capital availability does not appear to be a hindrance to the establishment of manufacturing within the city, neither do there appear to be any particular capital incentives to do so. Local tax rates and structures have been pinpointed as a problem by many local entrepreneurs and perhaps act as a disincentive for the establishment of industry. However, such a statement could probably be made about most locations and there is no available evidence that it is especially so in the case of Brandon.

*Markets*

Local markets are important for Brandon's industries. Confirmation of this can be found in the fact that the majority of the entrepreneurs in the city are themselves Brandon natives who have either inherited or established a concern serving local markets. It is worth restating that it is Brandon's situation rather than any inherent advantage of site or raw material base which has been, and continues to be, the major incentive for establishing manufacturing concerns.

Firms such as Ayerst Organics, Simplot Chemicals, and Inventronics have national and international markets and their addition to Brandon's manufacturing scene has been extremely important. Nevertheless, over the long term it has been the locally owned, locally developed, and locally run small manufacturing concerns which have provided the backbone of the slow but stready growth of the city's manufacturing sector. This trend is likely to continue.

Since Brandon possesses no unique features to attract industry, the function and leadership provided by the Brandon Industrial Commission has been of great importance. Responsible for the industrial base within the city, the current commission was established only in 1979, although similar bodies have existed for more than sixty years. While its main purpose is to attract and maintain industry within the city, industrial development is viewed as only one of the basic tools which, along with commercial, educational, recreational and residential development, will enhance the quality of life within the Brandon area.

The Brandon Industrial Commission strives to assist in the overall economic growth of industry and commerce in the greater Brandon community through

> Expanding and strengthening established industries. Efforts in this regard include acting as a clearing house and referral centre for services of senior governments. It also attempts to aid in resolving problems between firms and any levels of government. The Commission also refers requests, enquiries and ideas to other agencies in the community which can more appropriately deal with them.

Attracting new industries through maintenance of liaison with federal and provincial government departments; stimulating suggestions and assistance from local resources; doing selective advertising and providing immediate follow up to all potential developers.

Developing and maintaining relevant industrial and commercial statistics.[15]

The commission is much more interested in building upon the locally established base than it is in attempting to attract totally new industries, for linked industries are much more likely to provide a long-term stable environment. However, this does not preclude attempts to attract any industry which fits into the overall plan of development. Being pragmatic, however, the commission recognizes that western Manitoba does not have the same industrial attraction as do the industrial agglomerations of southern Ontario.

Most of Brandon's light industry is located along major access routes, a situation typical of most North American urban centres. The establishment of the Green Acres Industrial Park (Figure 4:5) with all major services is the city's attempt to control the location of all industry within its boundaries. Of major concern to the city administration has been the lack of road access to the industrial park without passing through residential areas. A provincial announcement of August 1985 promised such an access at Forty-Ninth Street East within the next four years.

## Conclusion

Since the city's beginning in 1881, Brandon's manufacturing sector has been an important, but not a dominant, feature in its growth. Tied to its agricultural base either as a source of raw materials, or as a source of markets, manufacturing has fluctuated with that base. Agricultural prosperity within the hinterland results in manufacturing prosperity within the city. The converse is also true. Concerted efforts over the years to broaden and diversify this base have met with little success. With all its advantages, Brandon has one major disadvantage that is impossible to overcome, the distance to major national and international markets. While Brandon may indeed be at the centre of Canada in the geographic sense, as the city's publication proudly claims, it is a fact of Canadian geography that Canada's major national markets are sixteen hundred kilometres away. Most industries cannot absorb the high transport cost.

The introduction of jet service to the city increases the possibility that it may be able to attract the more mobile "high tech" industries but this has not yet occurred. In light of historical evidence and the present situation, in the forseeable future expansion in the manufacturing sector in Brandon will remain slow and will continue to be based largely on local initiatives, supplies and markets.

**Notes**

1 See *Brandon, Manitoba, The Place to do Business* (Brandon: Brandon Industrial Commission, 1985).
2 Untitled document attributed to the city of Brandon, 1913.
3 Ibid.
4 *Trade and Commerce* (Brandon, September 1984).
5 Speech delivered by Mr. Reg Forbes, Brandon Industrial Commissioner, to Shur-Gro meeting, 6 March 1985. Mr. Forbes resigned as industrial commissioner in the fall of 1986.
6 *Brandon, Manitoba, The Place to do Business* (Brandon: Brandon Industrial Commission, 1985). Firms employing fewer than 10 people were not listed.
7 A difficulty that arises from attempting to derive time series statistics from DBS publications is the constant changing of categories from census to census. For the purposes at hand, however, the statistics being derived are only indicators and are not being subjected to powerful analytical techniques for comparison one against another. Because each evaluation is based entirely on the year in question it is felt that the numbers produced are both valid and useful.
8 The location quotient is derived using the formula:

$$\frac{\frac{ei}{et}}{\frac{EI}{ET}}$$

where: ei represents the local labour force engaged in a particular subgroup of economic activity (e.g., food and beverages);

et represents the local employment in a major industrial grouping (e.g., manufacturing);

EI represents the national employment in the sub-group;

ET represents the national employment in the major group.

9 *Trade and Commerce* (Brandon, September 1984).

10 *West-Man Transportation Study* (Brandon: Westman Regional Development Inc., 1969).

11 It is doubtful that all of the tonnage listed as originating in Winnipeg actually did so. As the principal break-in-bulk and warehousing point of the Canadian Prairies it is probable that much of the road freight comes from elsewhere. The study acknowledges this point when it states that the survey covered only the final distinct transportation service rendered.

12 Personal interview with Mr. Reg Forbes, March 1985.

13 *Background Studies, Brandon Development Plan* (Brandon: Province of Manitoba, Municipal Planning Branch, 1976).

14 Personal interview with Mr. Reg Forbes, March 1985.

15 Information provided by Mr. Reg Forbes concerning the Brandon Industrial Commission, August 1985.

## References

Fleming, S. 1879. *Report in Reference to the Canadian Pacific Railway*. Ottawa: MacLean Roger.

Smith, D.M. 1971. *Industrial Location: An Economic Geographical Analysis*. New York: John Wiley.

Trotter, B. 1925. *A Horseman and the West*. Toronto: Macmillan.

# SEVEN
# Downtown Brandon

*Christoph Stadel and John Everitt*

## Introduction

The city centre has been referred to by a variety of terms each having a slightly different semantic and spatial meaning. There is little unanimity on the precise definition and delimitation of the urban core. This is particularly true for the two terms commonly used by geographers — downtown and CBD. The downtown area has been described, often in a rather generalized fashion with little attempt being made at arriving at a precise delimitation, as a specific physical entity in the urban core, as well as the economic and social centre of the city and its major service area. The downtown area also offers the richest mixture of the past and present as well as displaying an indication of the directions for the urban future.

The term "CBD," although often used synonymously with downtown, is more clearly defined in the geographic literature:

> It was seen as the functional core of the city towards which the urban activities which required an accessible location for their economic viability or functional efficiency tended to gravitate. Typically, it comprised concentrations of retailing and associated consumer services, commercial and public office activities, wholesaling and warehousing and an array of entertainment functions such as theatres, hotels and cultural activities. It was suggested that these concentrations were arranged around the point of maximum accessibility to the surrounding hinterland and this usually coincided with the focus of the intracity transportation network. The concentrations of functions normally generated the highest vehicular and pedestrian traffic flows in the city and the highest density of daytime population. Competition for land to exploit the commercial advantages of accessibility produced the city's highest land values per unit area. Consequently, land use in this area was the most intensive in the city and this was

> reflected in the highest concentration of multistorey buildings. (Herbert and Thomas 1982:200)

Based on descriptive definitions of the CBD, attempts have been made to recognize variations in land use, and in the intensity of land use, by distinguishing specific sub-areas and functional clusters within the CBD. The most widely cited subdivision is the CBD core, CBD frame concept, originally used by Murphy and Vance (1954) in their attempt to arrive at an objective standardized method, the Central Business Index (CBI) method, to delimit the CBD. The method assumes a central point of maximum accessibility and highest land value in the centre of the CBD, the Peak Land Value Intersection (PLVI), on which the CBD core is focussed.

The definitions, the concepts, and the delimitation of downtown and CBD are mostly applicable to metropolitan centres although some common denominators can be found with smaller urban centres. Even in a small town, "Main Street . . . is probably its most remembered part. The visually most dominant buildings are located there and it is usually the social center of the town" (Hodge and Qadeer 1983:155). "Downtown Brandon" has been optimistically characterized as the "social and economic hub of the City . . ., the shopping center [and] the focus of business and civic life, not just for the inhabitants of Brandon itself, but for a large surrounding agricultural area, the most vital piece of real estate for many hundreds of square miles" (Underwood, McLellan and Associates 1967:69). In a study, *Brandon Downtown*, prepared by the Municipal Planning Branch of the province of Manitoba, the importance of the downtown is stressed as a "major public capital investment in utilities and services, the site of major private investment, a major source of public revenue through both business and property taxation, [and] perhaps the most visible feature of the City [contributing] immeasurably to Brandon's image as a regional service centre and as a potential urban-industrial centre" (Province of Manitoba 1971:6). Conversely, "Downtown Brandon" is "showing signs of decay and obsolescence which reduce its effectiveness and threaten to stifle its future growth and very existence" (Underwood, McLelland and Associates 1967:69). This problem of the "ugly conditions" (Hodge and Qadeer 1983:157) of downtown was unquestionably a contributing factor in the emergence of two suburban malls in Brandon during the early 1970s. As for cities in general, it appears that in Brandon the terms downtown and CBD have been used interchangeably and, as in the case of many other prairie cities, the downtown area of Brandon developed in conjunction with, and adjacent to, the main railway line which provided the initial impetus for urban development (Chapter 4).

Delimitation of Brandon's downtown and CBD in the various studies conducted by consulting firms and the Municipal Planning Branch has been done on a subjective basis. In *Background Studies: Brandon Development Plan* (Province of Manitoba 1977) it is stated that the "boundaries of the CBD consisting of 70 blocks were established on the basis of surveys of existing land and building uses within the city." Within the CBD, a commercial core of twenty blocks was designated as being "distinguishable from the remainder of the CBD by the type and quality of retail and business functions" (p. 2-43). Around the commercial core, several transitional zones of mixed zoning and land use were identified (industrial/commercial, commercial/residential, residental), the limits of the commercial core being established by the following procedure:

> To determine the limits of the Core area, floor space occupied by commercial uses was categorized on a block by block basis in accordance with the Standard Industrial Classification (S.I.C.). Those uses considered to be Core-commercial in nature included retail, financial, personal service and governmental activities. Non-Core commercial uses included wholesale trade, manufacturing activities, storage and garage facilities. Other uses such as transportation were classified Core-commercial or non-Core-commercial according to either their location vis-a-vis adjacent uses that could definitely be characterized as non-Core or their particular type of services. . . . Although high-density residential land and buildings may be considered a CBD use, and low-density a non-CBD use, they were excluded from the calculation of floor space existing in the CBD. (Province of Manitoba 1977:2-42)

Table 7:1 summarizes the nonresidential floor space in the core and transitional zones of the CBD of Brandon according to core and non-core use.

The CBD takes the form of a rectangle south of the CPR mainline with an extension to the south along the commercial ribbon of

TABLE 7:1
NONRESIDENTIAL FLOOR SPACE WITHIN THE CENTRAL BUSINESS DISTRICT OF BRANDON

| Type of use | Total CBD | | Core of CBD | | Transitional Zone of CBD | |
|---|---|---|---|---|---|---|
| | Sq. Ft. | % | Sq. Ft. | % | Sq. Ft. | % |
| Core Uses | 1,364,440 | 65.1 | 1,152,990 | 72.6 | 211,450 | 41.8 |
| Non-Core Uses | 160,700 | 7.7 | 147,400 | 9.3 | 13,300 | 2.6 |
| Vacant | 569,750 | 27.2 | 288,400 | 18.1 | 281,350 | 55.6 |
| Total | 2,094,890 | 100.0 | 1,588,790 | 100.0 | 506,100 | 100.0 |

*Source:* Based on a survey conducted June 1975 by the Municipal Planning Branch (Province of Manitoba 1977:2-43).

Tenth Street (Figure 7:1). In an attempt to apply the CBI method of Murphy and Vance (1954) to Brandon, Huff (1974) demonstrated the problems of applying this method to smaller urban centres. The problems are associated with the very modest vertical expression of the downtown skyline (Plate 7:1); the prevailing neglect of upper floor land use with high vacancy rates; less land use competition with, consequently, a greater land use mix; and the substantial proportion of downtown land used for parking lots. In his study Huff found that according to the CBI method, only seven city blocks, located between Pacific Avenue and Princess Avenue, would qualify as "CBD core" with one additional block as "CBD frame." For the Central Business Height Index,[1] the CBD blocks exceed only slightly the threshold index of one (highest index: 1.48). However, for the Central Business Intensity Index, which measures CBD-related land use, the indices exceed the threshold of 50 percent more substantially (highest index: 85.6 percent).

The desirability of downtown land and the related type and intensity of land use can also be derived from land value data. However, a number of problems may arise in working with land values: what type of land value should be used, how can access to often confidential data be obtained, and how realistic are the estimated market values or the assessed values?

In his paper on delimiting Downtown Brandon, Cowie (1975) used the more readily available assessed front-foot values adjusted to a standard depth of 120 feet, fully realizing that appraised land values would have been preferable for the purpose of determining the extent of the CBD. Using the PLVI (Tenth Street and Rosser

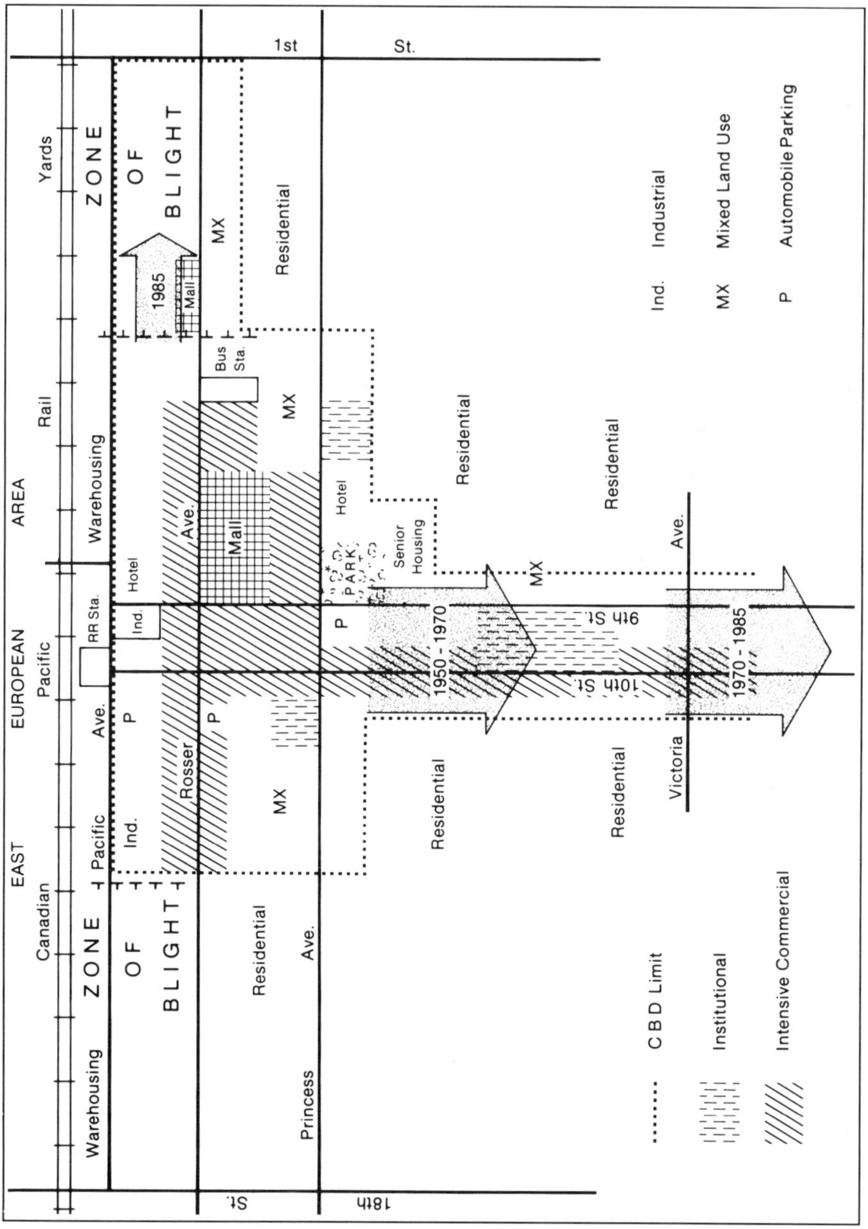


FIGURE 7:1

*Schematic representation of the Central Business District (CBD) of Brandon*

PLATE 7:1

*A recent (circa 1980) shot of Downtown Brandon from the northwest, with the Industrial Park in the background. The Central Business District is important commercially, although parking lots are as common as buildings in some areas. The rooftop of the Brandon Gallery dominates the centre of the picture with the McKenzie Seeds building and Scotia Towers nearer to the viewer also providing locational indicators. The Cargill elevators, on Pacific Avenue in the foreground, are located on the site of the old Great Northern station. The CPR tracks separate the downtown area from Assiniboine Gardens, north of the tracks, in the bottom left hand corner of the photograph. (Dirk Aberson)*

Avenue) as the focal centre of the CBD, he delimited the CBD core as including assessed front-foot values which reach at least 40 percent of those found at the PLVI, and the CBD frame as including values at least 20 percent of those at the PLVI. The mapped results of this study indicate that the area bounded by Rosser Avenue and Princess Avenue and by Seventh Street and Tenth Street form the CBD core. The CBD frame extends southwards along Ninth and Tenth Streets to Victoria Avenue and northwards to Pacific Avenue along Ninth Street. To the west and east, Rosser Avenue and Princess Avenue act as the major arteries for the extension of the CBD frame by about two blocks in each direction.

Cowie further proposed a combined method using both the assessed land values mentioned above and land use data (CBD land use versus non-CBD land use as defined by Murphy and Vance (1954)) to propose a CBD extending along Rosser Avenue and Princess Avenue between Fourth Street and Twelfth Street and between Pacific Avenue and Victoria Avenue along Ninth and Tenth Streets (Figure 7:1).

## Land Use Patterns — Continuity and Change

As the definitions in the previous section indicate, the terms downtown and CBD in North American cities are linked to "people, parcels, paperwork" (Yeates and Garner 1980:334), and to the functions of retailing, finance, professional services and offices, entertainment, and transportation. These are functions which require maximum accessibility and which experience daytime population flows.

Often, in large cities, distinct sub-areas of specific functions and land use can be distinguished with a marked gradient in land use intensity away from the PLVI. In smaller cities the size of Brandon, the distinctions are blurred and the typical CBD functions and land uses are mixed and blended with non-CBD uses. Given the lower value of land and a slower pace of speculation, outbidding, and development, the commercial cores of smaller cities not only show very modest vertical development of buildings and commercial activities and a less compact core, but also historic buildings, some of which date back to the foundation period of the city, may be preserved: "the lack of dramatic growth in small places, bemoaned by some, is also the reason that many older buildings still remain" (Hodge and Qadeer 1983:157). Although a large number of heritage buildings have been lost in Downtown Brandon due to neglect, fires, and the "bulldozer approach" to urban development, the city still exhibits a downtown architectural heritage that is worth preserving.

As would be expected, the CBD of Brandon contains the highest concentration of social, cultural, financial, professional and business activities in the city:

> Its [the CBD's] social, economic and administrative roles are particularly vital; it is the major centre of employment and trade within Brandon and the focus of commercial

> activities for the local, regional and community markets which it serves. Moreover, the CBD contributes immeasurably to the overall development of the community by conveying an image of Brandon's function as a regional service center and as a potential urban-industrial center. (Province of Manitoba 1977:2-42)

In *Background Studies*, it is also stated that the CBD is particularly important as an area of both public investment, in utilities and services, and private investment. The major functions of the CBD are summarized under four headings: (1) social and cultural functions; (2) governmental and institutional functions; (3) commercial function; and (4) wholesaling and manufacturing functions.

In terms of social and cultural functions, it is somewhat euphorically stated that the "CBD is the social and entertainment center of Brandon," and that it is set apart as "a special place for human interaction" and has "an air of excitement and interest" (p. 2-42). It is true that the CBD houses the new Arts and Library Complex, the YMCA, a spa, two movie theatres, lounges, restaurants and bars; but the social/cultural lead of Downtown Brandon is challenged by a number of establishments outside the area, mainly the Sportsplex north of the Assiniboine River, and Brandon University, the Westman Centennial Auditorium, Kinsmen Stadium, and the Keystone Centre along Eighteenth Street. In general, Eighteenth Street appears to be developing into a commercial ribbon with restaurants, taverns and lounges benefitting from good accessibility and parking. The social/cultural function of Downtown Brandon also appears to be curtailed by the fact that it contains a low number of permanent residents, with the major exception of senior citizens. It is surprising that in *Background Studies* the residential function is not even mentioned as one of the CBD functions. It can be argued that the present zoning regulations which classify a large area of Downtown Brandon as CC (Commercial Core) or GC (General Commercial) are a deterrent to the strengthening of the residential function.

Brandon's role as an administrative centre for the three government levels is reflected in the downtown land use. Governmental and institutional functions are concentrated mainly at the southern periphery of the downtown area, the City Hall and the Provincial Building both serving as "desirable anchors and major barriers to further CBD expansion to the south" (Province of Manitoba 1977:2-42).

The CBD remains the predominant commercial area of Brandon (Figure 7:2). Its commercial function encompasses a wide range of goods and services provided both for the population of Brandon as well as for the city's service area. The most prominent functional

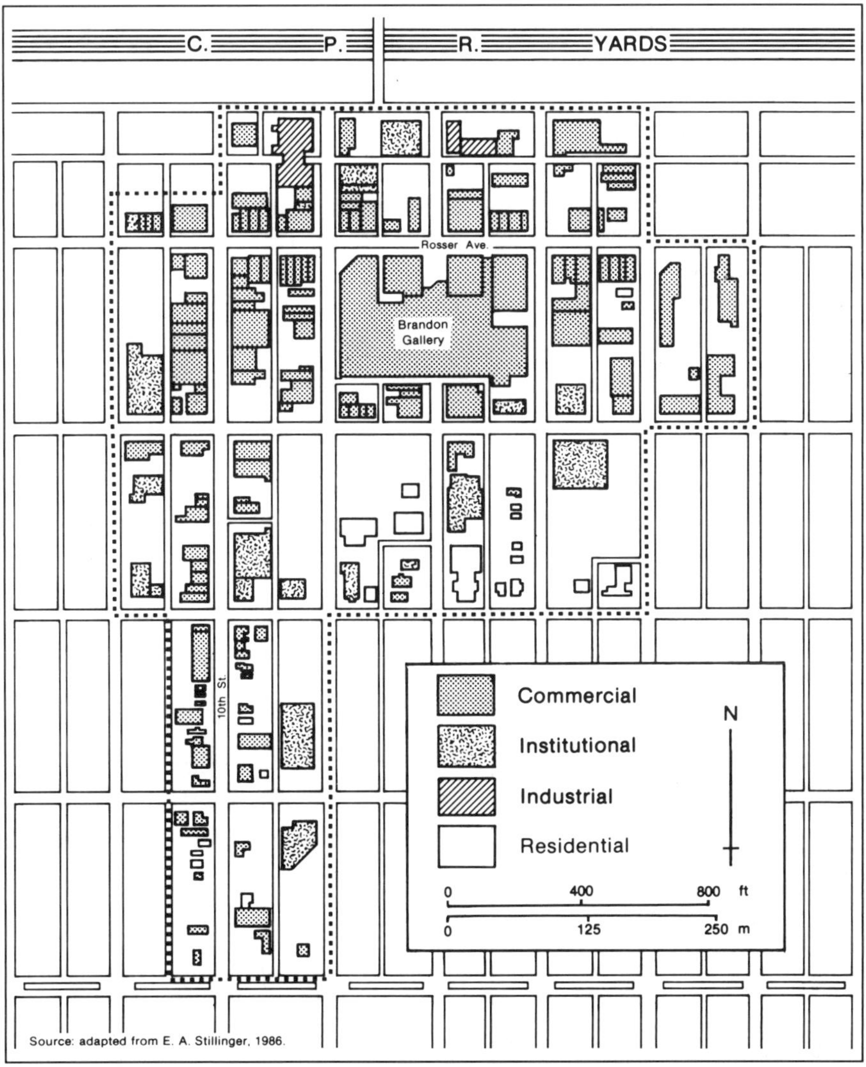


FIGURE 7:2

*Land use in Downtown Brandon in 1986 (the dashed line delimits the Central Business District)*

categories in Brandon's commercial core are: department stores (Eaton's, Zeller's, Stylerite, Sears, Metropolitan); specialist retail outlets (e.g., food, clothing, dry goods and shoe stores; drug stores and photography stores; sporting goods stores; book and stationery stores; jewellers; hardware, gardening and furniture stores; optical and hearing aid stores; music stores; flower shops; and gift and craft shops); motor vehicle dealers and automotive services outlets; gas stations; specialized services (e.g., travel agencies, barber shops and beauty shops); financial services (banks, trust companies, credit and investment agencies, insurance companies and real estate companies); and professional services (e.g., lawyers, accountants, engineers and clinics). *Background Studies* (Province of Manitoba 1977:2-79, 2-80) lists, a total of 247 commercial establishments in the CBD of which 150 are classified as retail outlets, 63 as financial outlets, and 134 as services. This compares with 139 retail outlets, 19 financial establishments, and 85 services outside the CBD. The commercial function, especially the retail trade and services, is concentrated along the traditional street ribbons of Rosser Avenue, Princess Avenue, Ninth Street and Tenth Street as well as in an enclosed downtown mall (Plates 7:2 and 7:3). The latter, known as the Brandon Gallery, has approximately 22,760 square metres of retail space and a total of 51 stores and services anchored by the two department stores, Eaton's and Zeller's (The Bay until June 1985).

The commercial function of the CBD has encountered strong competition from the suburban Brandon Shoppers' Mall (established in 1971, with an annex being completed in 1974) and from the Canadian Tire store (opened in 1973). In 1981, the Brandon Shoppers' Mall expanded to cover 36,045 square metres with a total of seventy-three stores and services, anchored by Woolco and Safeway. With this major commercial development at the intersection of Eighteenth Street and Richmond Avenue, the west side of Eighteenth Street rapidly developed into a commercial ribbon, especially opposite the Keystone Centre and the Provincial Exhibition Grounds located on the east side of Eighteenth Street (Plate 7:4). In the early 1980s, a new mini street-mall, the Thomas Mall, which was expanded in 1985 to a potential total of 23 establishments and a gross floor area of 2,508 square metres, further strengthened the commercial role of Eighteenth Street. The ribbon commercial development along Eighteenth Street recently spilled

PLATE 7:2

*Rosser Avenue, the major retail artery of Downtown Brandon, is flanked by flat-roofed commercial blocks — to the left the old Cement Block (1910), to the right the new downtown mall, the Brandon Gallery (1980). In the background are the well-treed older residential areas of the city. Manitoba Hydro's thermal electric plant can be seen on the skyline. (John Welsted, 1987)*

PLATE 7:3

*Tenth Street is the major commercial artery extending southward from Downtown Brandon. In the foreground are some of the older retail establishments; in the background (farther south) is the newer, post-1950, extension of the commercial core. The tall buildings on the left side are the Princess Towers, a senior citizens' housing complex of the Manitoba Housing Corporation, located on the site of the fomer city hall. In the background are the well-treed older residential areas with the gentle arc of the Brandon Hills on the skyline. (John Welsted, 1987)*

northwards across the railway bridge onto the Assiniboine floodplain with the establishment of automotive services, three new hotels, two restaurants, and another mini-mall, the Redwood Mall, which was established in 1983 and expanded in 1984 to include seven retail outlets and a gas station. The mini-malls are classified in *City of Brandon Development Plan* (Province of Manitoba 1977) as "community commercial" establishments with a maximum allowable gross floor area of 30,000 square feet (2,787 square metres) and a minimum distance of one mile (1.6 kilometres) from another regional or community commercial district, the latter requirement subsequently abolished by city council. Downtown Brandon experienced further erosion of its commercial role by developments in the western suburbs, especially with the establishment, in 1971, of the K-Mart/IGA (formerly Locomart) complex at the intersection of Victoria Avenue and Thirty-Fourth Street. With the establishment of this shopping centre and rapid residential development in the west

PLATE 7:4
*View north along Eighteenth Street with the junction of Eighteenth Street and Richmond Avenue in the middle of the photograph. The Brandon Shoppers' Mall is located southwest (bottom left) of the junction, with the Keystone Centre complex to the northeast. The CNR line cuts through the centre of the photograph and Downtown Brandon can be seen in the top right hand corner. Compare this photograph with Plate 6:1 taken in 1960. (John Everitt, 1987)*

end of Brandon, Victoria Avenue developed into a commercial ribbon with automotive services, two hotels, restaurants, and other retail outlets and services.

It is interesting to note that *City of Brandon Development Plan* clearly states that "it is essential, insofar as possible, [that] most of the major commercial and business developments be directed to the central core" and that "no new community or regional shopping centres, as distinct from neighbourhood shopping facilities, should be established elsewhere in Brandon for at least the next 15 to 20 years" (Province of Manitoba 1977:3-9, 3-10). It appears that this objective has been only partially achieved at best and that, although no new regional shopping centre has been established, a major expansion of an existing one, the creation of a number of mini-malls, and a further expansion of commercial ribbons has been allowed.

The wholesaling and manufacturing functions have been traditionally associated with the fringe of CBD areas for locational and transportation reasons. It is particularly the northern fringe of the CBD towards the CPR mainline along Pacific Avenue that exhibits a concentration of these functions (Plate 7:5). A study carried out by Gibbons (1983) on the upper-floor land use of Downtown Brandon found that wholesale and manufacturing occupies a significant proportion of the upper-floor space at the northern edge of the CBD. *Background Studies* (Province of Manitoba 1977:2-42, 2-43) reports that thirty-three wholesale trade and thirty manufacturing industry establishments are found in the CBD, representing approximately one-third of the total number of wholesale and manufacturing establishments in the city. As the locational and transportation advantages near the commercial core and the railway line are no longer as great in the age of truck-

PLATE 7:5

*Brandon owes its existence to the transcontinental railway which reached the area in 1881. To the south of the railway (to the left on the photograph) a grain-handling and wholesaling ribbon developed along Pacific Avenue. Further south the photograph shows the haphazard structure of the western fringe of Downtown Brandon. In the background are the western residential areas of the city with the floodplain of the Assiniboine River north of the railway tracks. The slight dip on the horizon is the valley of the Assiniboine River. (John Welsted, 1987)* oriented transportation as they were during the railway era, and, as the new objectives of planning and urban policy prefer concentration of these functions at the eastern periphery of the city in the Green Acres Industrial Park, the traditional wholesale and manufacturing district has been affected by functional obsolescence and structural blight.

Among the other functions and types of land use in Downtown Brandon, although they are not specifically mentioned in *City of Brandon Development Plan*, are transportation-related land use (streets, lanes, parking lots, CPR station and yard, Greyhound and Grey Goose bus depots), residential land use, recreational land use, and vacant space (Table 7:2). Although residential land use is generally considered a non-CBD activity and permanent nighttime population is almost totally absent from the CBDs of many large North American cities, Brandon's CBD still contains a significant residential population, generally either on the upper floors of commercial buildings in the core or in older houses at the periphery

TABLE 7:2
LAND USES WITHIN THE CENTRAL BUSINESS DISTRICT OF BRANDON

| Types of Use | Total CBD | | Core of CBD | | Transitional Zone of CBD | |
|---|---|---|---|---|---|---|
| | Acres | % | Acres | % | Acres | % |
| Residential* | 56.8 | 23.4 | 5.0 | 6.4 | 51.7 | 31.5 |
| Hotel | 1.8 | 0.8 | 1.8 | 2.3 | 0 | 0 |
| Commercial | 44.1 | 18.2 | 18.5 | 23.6 | 25.6 | 15.6 |
| Industrial | 5.3 | 2.2 | 0.4 | 0.5 | 4.9 | 3.0 |
| Transportation & Utilities | 8.7 | 3.6 | 2.3 | 2.9 | 6.4 | 3.9 |
| Parks & Recreation | 4.1 | 1.7 | 1.1 | 1.4 | 3.0 | 1.8 |
| Government | 3.8 | 1.6 | 3.7 | 4.7 | 0.1 | 0.1 |
| Churches, Halls & Clubs | 2.8 | 1.2 | 1.5 | 1.9 | 1.3 | 0.8 |
| Parking | 26.7 | 11.0 | 16.3 | 20.8 | 10.4 | 6.3 |
| Vacant | 4.5 | 1.9 | 0.5 | 0.6 | 4.0 | 2.4 |
| Rights-of-way | 84.2 | 34.7 | 27.2 | 34.8 | 57.0 | 34.7 |
| Total | 242.7 | 100.0 | 78.3 | 100.0 | 164.4 | 100.0 |

*This includes only sites on which residential buildings have been constructed. It excludes residential suites in the upper storeys of commercial buildings.

*Source: Background Studies: Brandon Development Plan* (Province of Manitoba 1976:2-44).

of the CBD. Furthermore, it has been a planning objective to retain, or even to increase, residential population in the downtown area as a means of revitalizing the core. The Brandon Housing Authority, under the auspices of the Manitoba Housing and Renewal Corporation, has built three high-rises for senior citizens in the core of the downtown area. These have a total of 224 dwelling units, and another two apartment blocks in the CBD transitional area have 130 dwelling units. In 1985, the zoning bylaws were changed to allow residential land use in the downtown area on and above the second floor. Kwiatkowski (1984) in a study, *Housing in Downtown Brandon*, found 167 dwelling units in 45 different buildings in the commercial core (CCC designation according to the City of Brandon Zoning Bylaw of 1982). These are concentrated at the southern periphery of the commercial core while four entire blocks had no residential land use at all. The largest concentration of dwellings (36 units) is found on the upper five floors of the Scotia Towers, at the corner of Rosser Avenue and Tenth Street, with another 46 dwellings being located along Tenth Street between Rosser Avenue and Princess Avenue (second and third storey apartments in older buildings). Whereas housing in the core of the CBD is generally found on the upper floors of commercial buildings, there tends to be a mixture of residential and commercial buildings at the southern periphery of the CBD core. Particularly along Tenth Street between Princess Avenue and Victoria Avenue, a number of former residential dwellings have been replaced by, or converted to, commercial buildings; in some cases a commercially used addition has been attached to the residential dwelling.

In comparing residential densities within the CBDs of large urban centres and the small city of Brandon, one finds that the highest densities in Brandon are at, or close to, the PLVI (Rosser Avenue and Tenth Street) in contrast to many North American metropolitan centres where residential densities tend to be lowest around the PLVI. In terms of the socio-economic characteristics of the residents, Kwiatkowski found an over-representation of young adults and retired people and a high proportion of residents with low incomes and no permanent employment. This, however, does not apply to the population group living in the expensive suites of the Scotia Towers.

An interesting aspect of Brandon's downtown land use relates to the use of the upper floor space. As indicated earlier, the CBD of large urban centres exhibits a distinct vertical zoning of land use in high-rise buildings. As the competition for commercial space is far less intensive in smaller centres, this "vertical component" is largely missing in the case of Brandon, although the planning concept for the city favours the "Scotia Towers" model with the lower levels being devoted to banking, retailing and professional services, and the top levels to residential use. However, in Brandon's downtown the upper floor spaces generally are poorly used or vacant and are often in a stage of blight.

By comparing the situation documented in *City of Brandon — Urban Renewal Study* (Underwood, McLellan and Associates 1967) with that of 1978 described in a study by Carter (1978), and that outlined most recently by Gibbons (1983), one notes that this problem has persisted throughout the last few decades. The upper floors are used predominantly for inexpensive (except for Scotia Towers), often substandard, residential suites; manufacturing; wholesaling and storage (often in a condition of blight); inexpensive (except for the Canadian Inn) hotel rooms; offices; and financial services and government functions in more modern or upgraded structures. The inefficiency of upper floor space use is closely related to the age of buildings in the commercial core, many structures dating back to the pre-World War II or even the pre-World War I period, and a failure to maintain the building fabric properly.

In addition to the structural deterioration and blight, and the functional obsolescence of buildings, the inability or unwillingness to adapt them to modern requirements and safety regulations has to be mentioned. The latest example of this was the Metropolitan Department Store which occupies the same city block as the Brandon Gallery but which, oddly enough, is not functionally integrated into the mall (Plate 7:2). It was one of the prime examples of blight in the commercial core. When the Brandon Gallery was built, the windows of the upper two floors of the store were filled in with bricks and the interior was more or less left to deteriorate to the point that fire regulations were no longer met. After several warnings, the Metropolitan Store was given a last period of grace to either upgrade the condition of the upper floors or to remove them. Finally, in April 1986, the upper two floors of the store were removed, and the last commercial building of the 1880s in its original form disappeared from the downtown area.

Given these kinds of conditions it is not surprising that the downtown has experienced several fires during the last few decades, destroying a number of old structures which have been only partially replaced by new buildings.

In the selection and spatial arrangement of sites within the CBD, a number of principles, structural models and activity linkages have been suggested in the literature (Rannels 1956; Nelson 1958; Horwood and Boyce 1959; Ward 1966; Davies 1972; Davies and Bennison 1978; Hartshorn 1980; Herbert and Thomas 1982). It appears doubtful, though, that these can be easily applied to the situation of a small prairie city like Brandon. Even for larger urban centres, it has been pointed out that the "vagaries of the local land market and the historical process of development introduces further complexity into the spatial pattern of potential retail sites" and that "the detailed locational decisions both of businesses and individuals are influenced by a number of principles which seem unlikely to result in a simple geometrical arrangement of shop types" (Herbert and Thomas 1982:206).

An attempt has been made to provide a schematic representation of the CBD of Brandon (Figure 7:1). It depicts the T-shape of the CBD formed by the crosslike arrangement of the major commercial arteries of Rosser Avenue and Tenth Street, and it summarizes the major land use components of the CBD with the downtown mall (Brandon Gallery) as the principal commercial anchor. Finally, major growth directions of, and barriers to, the expansion of the CBD are shown.

## Problems of the Downtown Area and the Planning Response

As early as 1967 it was recognized that "decay and the need for renewal affect the whole of the City fabric but it is in the Central Business District that the problem is most serious and most difficult to solve" and that "the revitalization of the downtown area is the most important single problem facing the citizens of Brandon" (Underwood, McLellan and Associates 1967:69). More specifically, *City of Brandon — Urban Renewal Study* mentioned that the components of the city centre had become old, outgrown, out-of-date, and functionally obsolete.

In terms of transportation and transport-related land use and problems, the *Study* noted the prevailing tendency in most downtown areas of North American cities, at least until the 1970s, to "ease the flow of traffic at the expense of the convenience of the pedestrian" (p. 70), a tendency which could also be observed in Brandon. Yet it was recognized that the major function of the city centre should be that of a market place and meeting place, functions which could be stimulated by creating an environment for people on foot. However, "the universal car ownership which has made downtown inconvenient for the shopper is a flexible means of transportation and the ties to downtown by public transport and pedestrian accessibility have been weakened" (p. 70). The *Study* came up with the conclusion that "the provision for both long and short term parking spaces [in downtown Brandon] is inadequate, poorly distributed and an inefficient use of land" (p. 78). It was also observed that the provision of truck loading zones was inadequate, and that the rear servicing facilities through side lanes and back lanes are unsatisfactory and a source of visual blight.

The *Brandon Area Transportation Study* (City of Brandon 1970), examining the traffic and parking system, evaluated for the first time the possible effect of closing part of Rosser Avenue for vehicular traffic to create a downtown mall. Assuming that all east-west streets would become two-way streets, the study concluded that Princess Avenue would become "the major traffic route carrying volumes of approximately 20,000 per day" and that "all parking would have to be eliminated on Princess Street [*sic*]" but that "the idea of a downtown mall is not unreasonable from the traffic standpoint" (p. 52).

The first of a series of comprehensive redevelopment and downtown mall concepts was the Rosser Avenue Mall Project (RAMP), designed and formulated by architect Glenn W. Lawson (City of Brandon 1971). It proposed converting Rosser Avenue between Sixth and Tenth Streets into an enclosed mall with the roof structure becoming a major parking deck made accessible by a series of ramps from the south. However, this project did not get much beyond the stage of architectural design, and preliminary legal and financial considerations.

In 1971, shortly before the opening of the two regional shopping centres (the Brandon Shoppers' Mall and K-Mart), the *Brandon Downtown Preliminary Report* optimistically stated that "decentralization of commercial facilities has not yet reached a point where the Brandon Downtown has been irreversibly damaged" although it was already recognized that "the parking facilities, improved vehicular circulation, and the climate-controlled mall of the shopping centre development may weigh heavily against the Downtown merchant" (Province of Manitoba 1971:1). The *Report* goes on to brand the CBD as an "overzoned low-density [area], the retail core of which is threatened by functional obsolescence, inadequate vehicular circulation and parking, increasing competition from decentralized sources, and from the general lack of vitality and dynamics" (p. 1-2). It was also recognized that the initiative for redevelopment must come from both civic administration (through zoning and other measures) and from the private sector (mainly through a financial commitment).

A rather comprehensive study of Downtown Brandon, with an evaluation of various strategies and design concepts, was carried out in 1972 by Gordon Hlynsky in a thesis, "Strategy for Town Centre Development — Brandon," for the Faculty of Architecture, University of Manitoba. Three design concepts for a downtown mall were evaluated: (1) the Rosser Avenue Mall Project; (2) a core town centre concept with a climate-controlled mall and rehabilitated structure around it; and (3) a large multi-level mall structure between Rosser Avenue and Princess Avenue, a concept which was favoured by the authors of this chapter and which was subsequently adopted in a modified form with the eventual construction of the Brandon Gallery mall. Hlynsky, however, proposed a multi-functional mall with the integration of offices, a hotel, apartments, cultural facilites, and parking on lower levels. Although this concept of multi-functional malls has been adopted in many cities, the Brandon Gallery was limited to a retail/parking function. Although the Hlynsky study also recommended a revitalization of the area adjacent to the mall, no specific designs were proposed.

In 1975, the consulting firm of Damas and Smith was commissioned to propose a *Brandon Downtown Redevelopment Program* (Damas and Smith 1975). Among the problems mentioned was the inefficient use of land in Downtown Brandon with a considerable amount of surface parking, vacant land, and vacant building floor space. It was pointed out that this is not only visually unappealing but that it also means a loss of tax revenues. It was further observed that the downtown area has a "rather run-down and neglected appearance, [that] it seems to be very dispersed and spread out over a large area and [that] it does not seem to have any concentration of activities or focus which acts as a major attraction and which imports some of its magnetism to the rest of the downtown" (p. 2). It was also stated that these problems are not peculiar to Brandon and that many North American cities were in an even worse shape to the point where the downtown "has virtually disappeared as a definite and recognizable centre for the community. . . . Brandon's downtown, of course, is not anywhere near that condition of dereliction. In spite of the defects indicated it is still a central area of considerable viability and potential" (p. 2).

The fact that social problems appear to be concentrated in the central area tends to represent a constraint for redevelopment and revitalization. Finally it was observed that "the effect of the suburban shopping centres on Brandon's downtown has been felt but [that] it is not yet critical." However, it was considered that the addition of another suburban shopping centre would have a "seriously damaging effect" on the downtown and that

> The erosion of the downtown is regarded as undesirable, not only by those who have a special commercial interest in its preservation but also by many who believe that the special characteristics of a vital central area contribute something unique and attractive to the life of a city. (p. 5)

Therefore, the declared objective of the Damas and Smith study was "to formulate a strategy which will help to preserve Brandon's central area and to ensure its continued growth not only as a vigorous commercial centre, but as a healthy and flourishing central part of the entire community where its citizens can find the best kinds of urban experiences" (p. 5). The approach adopted to achieve these goals was called an "action orientation"; that is, identification of those development projects which were most feasible, the formulation of the conceptual development framework for those projects, and an attempt to undertake those projects by direct negotiations with potential developers.

The proposed "Master Project" called for the construction of a multi-level, multi-functional mall complex between Rosser Avenue and Princess Avenue and between Seventh and Ninth Streets, to include two levels of retail space, apartments, a hotel, the city's Arts Centre and underground parking. No developer could be found for this ambitious project. It had to be revised and scaled down but it remained the basic concept for the subsequent Brandon Gallery mall — without the Arts Centre, without the hotel and apartments, and with rooftop rather than underground parking.

More recent studies of the downtown area of Brandon were carried out by the Municipal Planning Branch of the province of Manitoba in connection with the *City of Brandon Development Plan* and the *Brandon and Area Planning District Development Plan* (Province of Manitoba 1984). In the *Background Studies: Brandon Development Plan*, the following problems of Downtown Brandon were identified and were restated in the *City of Brandon Development Plan*:

> (i) the excessive amount of vacant but unsuitable commercial floor space in the C.B.D.; (ii) lack of compactness and cohesion; (iii) lack of suitable major anchors to act as true magnets; (iv) the obsolescence evident in many of the older, multi-storey buildings which lack one special basic technology to make them more useful and useable, i.e. an elevator system; (v) the zoning by-law which encourages industrial uses in the C.B.D. whilst permitting C.B.D. uses in outlying commercial districts; (vi) value and assessment on these older buildings necessarily reflect their obsolescent condition. (Province of Manitoba 1977:3-9)

It was pointed out that these problems resulted in the following negative consequences for the CBD and the Brandon community as a whole:

> (i) loss of public capital investment in the C.B.D.'s utilities and services as they are not used to their full capacity and life expectancy; (ii) incurment of the costs of additional public capital investments for services and utilities in outlying commercial districts; (iii) loss of private capital investments because of declining C.B.D. building and land values; (iv) loss of public revenue through reduction of taxes for C.B.D. businesses and property; (v) deterioration of the City's image as a regional service centre and potential urban-industrial centre; (vi) lowering the C.B.D.'s attractiveness as a shopping centre and as a potential location for investment development; (vii) potential loss of jobs, if because of the general unattractiveness, the decline of the central area is accom-

panied by the inability of the City to draw new commercial development and manufacturing operations to other areas of the City. (Province of Manitoba: 3-9, 3-10)

In its proposals for the CBD, the *City of Brandon Development Plan* spells out a number of policies for the core which may be summarized as follows:

> (i) It should be restricted in size, with clear demarcation of its limits, and it should be upgraded while maintaining its multi-functional role. (ii) It should be preserved as the main pedestrian-oriented retail shopping area, and there should be a ban on the construction of new community or regional shopping centres outside the CBD. (iii) It should be zoned as a Central Commercial Restricted Area permitting only intensive core-type uses which should be prohibited outside the CBD. (iv) A physical beautification programme should be started aimed at creating a pleasant pedestrian-oriented image. (v) Historically important buildings should be protected. (vi) A second major magnet (apart from the downtown mall) should be created in the western part of the core. (vii) A "round-the-clock" population should be encouraged within the CBD by the provision of high density residential accommodation with a wide range of facilities and amenities. (viii) Government facilities and amenities should be encouraged to locate within the core. (ix) Adequate and suitable landscaped parking space should be provided. Parking space should be integrated into new facilities and on-street parking should be reduced gradually by the provision of parking structures at the periphery of the core. (x) Efforts should be made to eradicate the problem of vacant and decrepit floor space and underutilized land. (xi) A small portion of the business and assessment taxes should be designated for the purpose of implementing a number of downtown development projects. (xii) A downtown development committee, a consultative civic group, should be formed. (Province of Manitoba 1977: 3-10, 3-11, 3-12)

In order to implement these policies gradually, a five-stage development and redevelopment concept was suggested with stage one including the area of the two blocks presently occupied by the Brandon Gallery (Figure 7:2). Stage two, including the two blocks west of the Brandon Gallery, should be concluded by 1991. Stages three and four would attempt to rehabilitate and redevelop the core area blocks between Rosser Avenue and Pacific Avenue. Stage five would eventually be carried out in the areas adjacent to the eight blocks in the centre of the CBD. Finally, the plan included a conceptual long-range plan for the layout and functional composition of the core area.

As mentioned earlier, the downtown mall (Brandon Gallery), considered to be the essential part of stage one, was completed by the end of 1980. On the advice of a report by Levin, done while he was with Damas and Smith, the city assembled the land at a cost of $625,000 and is leasing it to the mall owners — Cambridge Leaseholds, Eaton Properties Ltd., and 80860 Canada Ltd. (major shareholder Damas and Smith) — for ninety-nine years at a cost of one dollar. In the words of the mayor this was an investment that was going to be "recouped quickly through increased tax revenue" (*Brandon Sun* 06/07/1982). Some new businesses were attracted to Brandon and the downtown mall; some established premises in both the Brandon Gallery and the suburban Brandon Shoppers' Mall, while others moved their operation from downtown stores to the downtown mall. Vacancy rates in the Brandon Gallery during the last five years have fluctuated between 0 and 10 percent, generally somewhat higher than in the suburban Brandon Shoppers' Mall. The initial economic success of the Brandon Gallery was hampered by a general economic recession in the early 1980s and by the $6.5 million expansion of the Brandon Shoppers' Mall in November 1981 which increased its size by approximately one-third and included a greatly enlarged Safeway as a major grocery supermarket. In contrast, the Brandon Gallery had major problems in finding a food retailer as a tenant. Economy Foods is now occupying a somewhat peripheral first-floor location adjacent to the deck parking but is much smaller than both Safeway outlets within the city.

As the expansion of the suburban mall was not a new regional shopping centre, it did not violate the "letter" of the *City of Brandon Development Plan*; however, it may have worked against its "spirit." Although the effects of this expansion on the Brandon Gallery and Downtown Brandon are not clear, it touched off considerable controversy. Former Alderman Borotsik, who at the time was also the manager of the Brandon Shoppers' Mall, declared that "if anything we have probably added to the Gallery by attracting shoppers from our market radius" (*Brandon Sun* 07/07/1982). Others, however, argued that it "fragmented shoppers' desires" (former Alderman Hudy) and that Bramalea Ltd., the Shoppers' Mall's former owners, "set out to purposely try to downgrade the supposed good image that was being planned for the Downtown" (Alderman Cayer). Former Alderman Dyck went as far as to say: "I looked upon the expansion on the same basis as if we had developed a new mall out there. By allowing it we certainly didn't give the support to the downtown that we were supposedly dedicated to" (*Brandon Sun* 07/07/1982). Nevertheless most people would agree that the Brandon Gallery halted the further decay and economic decline of Downtown Brandon and that it is a very important stage in the redevelopment of the commercial core. Borotsik, however, declared that "it [the Brandon Gallery] was supposed to turn Brandon into something that it never was and in my opinion never could be. If anything there are as many if not more retail holes [*sic*] than there were before" (*Brandon Sun* 05/07/1982).

In his Master's thesis, "Downtown Revitalization Strategy: Brandon, Manitoba," E.A. Stillinger (1986) critically assesses past studies of, and projects for, Downtown Brandon. He then proceeds to analyse the structure and physical functions of the downtown area, with the aid of a series of overlay maps and a photographic survey. In the last chapter of his thesis, Stillinger proposes a downtown revitalization strategy for Brandon, his general suggestions including: (a) the encouragement of traditional building and development type; (b) the renovation or rehabilitation of historic buildings to standards compatible with modern uses; (c) an emphasis on pedestrian presence; and (d) the improvement of a pedestrian environment (p. 61).

Since the mid-1980s, it appears that downtown businessmen have finally become more committed to revitalization of Brandon's core area. In 1985, the Downtown Brandon Association voted unanimously to take advantage of the Heritage Canada Mainstreet Program. On 16 December 1985, city council agreed to "allocate $200,000 to begin the process of improving the image and aesthetics of downtown streets" (*Mainstreet Brandon* 1986:1). The Mainstreet Canada Program, instituted in 1980, provides free assistance with store-front designs and helps find sources of funding for renovations. According to a former chairman of the Downtown Development Committee, "the Program does not supply the money for improvements or promotions but it does get the three levels of government as well as the businessmen involved in looking for ways of reversing the trend away from downtown" (*Brandon Sun* 20/06/1985). Furthermore, under the terms of the Heritage Canada Mainstreet Program, a full-time coordinator was hired. In late fall of 1986, developers approached city council to get permission for the construction of a new suburban mall in the vicinity of the Brandon Shoppers' Mall and to lift the moratorium which bans construction of suburban malls until 1994. Council was split on this issue but the development proposal stalled when it was revealed that the land option of the developer had expired. In the spring of 1987, another group of developers approached the city with a proposal to build a mall on the "North Hill." City council approved only a mini-mall up to a maximum of thirty-two thousand square feet; to date it is unclear whether the mall will be built.

On the other hand, action towards a revitalization of downtown Brandon has continued. In the spring of 1987, the consulting firm of Hilderman, Wilty and Associates was hired "to design a plan for the rebirth of the core area" (*Brandon Sun* 25/03/1987). The new chairman of the city's Downtown Development Board, Jeff Harwood, optimistically proclaimed that "the whole process of redevelopment and revitalization of the downtown has begun" (*Brandon Sun* 25/03/1987). City council also committed $350,000 a year for five years to downtown development providing that businessmen in the area agreed to form a Business Improvement Area. This would commit the businesses in the downtown area to paying a special levy for the upgrading and revitalization of the downtown. In 1987, city council approved the formation of a business improvement area for Downtown Brandon.

## Notes

1. Central Business Height Index (CBHI):

$$= \frac{\text{Total floor space of all central business land uses in the block}}{\text{Total ground floor space of the block}}$$

(Values greater than 1 are considered potentially indicative of the CBD.)

Central Business Intensity Index (CBII):

$$= \frac{\text{Total floor space of all central business land uses in the block}}{\text{Total floor space of all land uses in the block}} \times 100$$

(Values of greater than 50 percent are considered potentially indicative of the CBD.)

**References**

*Brandon Sun*, 5 July 1982, 6 July 1982, 7 July 1982, 20 June 1985, 25 March 1987.

Carter, S. 1978. A Study of Upper Floor Use in Downtown Brandon. Student paper, Department of Geography, Brandon University.

City of Brandon. 1970. *Brandon Area Transportation Study*. Brandon: City of Brandon.

City of Brandon. 1971. *Rosser Avenue Mall Project (RAMP)*. Brandon: City of Brandon, Industrial Commission.

Cowie, S. 1975. Delimiting Downtown Brandon. Student paper, Department of Geography, Brandon University.

Damas and Smith. 1975. *Brandon Downtown Redevelopment Program*. Brandon: City of Brandon.

Davies, R.L. 1972. Structural models of retail distribution: analysis with settlement and land-use theories. *Transactions, Institute of British Geographers* 57:59-82.

Davies, R.L. and D.J. Bennison. 1978. Retailing in the city centre: the character of shopping streets. *Tijdschrift voor Economische en Sociale Geografie* 69(5):270-85.

Gibbons, K. 1983. Upper Floor Land Use in Downtown Brandon. Student paper, Department of Geography, Brandon University.

Hartshorn, T.A. 1980. *Interpreting the City: An Urban Geography*. Toronto: John Wiley.

Herbert, D.T. and C.J. Thomas. 1982. *Urban Geography: a First Approach*. Toronto: John Wiley.

Hlynski, G. 1972. Strategy for Town Centre Development — Brandon. M.A. thesis, University of Manitoba.

Hodge, G.D. and M.A. Qadeer. 1983. *Towns and Villages in Canada: The Importance of Being Unimportant*. Toronto: Butterworths.

Horwood, E.M. and R.R. Boyce. 1959. *Studies of the Central Business District and Urban Freeway Development*. Seattle: University of Washington Press.

Huff, K. 1974. The Central Business Index Method Applied to Brandon, Manitoba. Student paper, Department of Geography, Brandon University.

Kwiatkowski, M.J. 1984. Housing in Downtown Brandon. Student paper, Department of Geography, Brandon University.

Mainstreet Brandon. 1986. 1(1):1.

Murphy, R.E and J.E. Vance. 1954, Delimiting the CBD. *Economic Geography* 30:197-223.

Nelson, R.L. 1958. *The Selection of Retail Locations.* Chicago: F.W. Dodge Corporation.

Province of Manitoba. 1971. *'Brandon Downtown Preliminary Report.* Brandon: Province of Manitoba, Municipal Planning Branch.

Province of Manitoba. 1976. *Background Studies: Brandon Development Plan.* Brandon: Province of Manitoba, Municipal Planning Branch.

Province of Manitoba. 1977. *City of Brandon Development Plan.* Brandon: Province of Manitoba, Municipal Planning Branch.

Province of Manitoba. 1984. *Brandon and Area Planning District Development Plan.* Brandon: Province of Manitoba, Municipal Planning Branch.

Rannells, J. 1956. *The Core of the City.* New York: Columbia University Press.

Stillinger, E.A. 1986. Downtown Revitalization Strategy: Brandon, Manitoba. M.A. thesis, University of Manitoba.

Underwood McLellan and Associates. 1967. *City of Brandon — Urban Renewal Study.* Winnipeg: Underwood McLellan and Associates.

Ward, D. 1966. The industrial revolution and the emergence of Boston's Central Business District. *Economic Geography* 42:152-71.

Yeates, M. and B. Garner. 1980. *The North American City.* San Francisco: Harper and Row.

# EIGHT
# The Urban Fringe

*John Everitt and Christoph Stadel*

## Introduction

Although Brandon is a relatively small city in the Canadian context (Pearson 1970), it is the dominant urban centre in southwest Manitoba and consequently has a major influence on its hinterland (Lewis and Maund 1976) and is in turn strongly influenced by the region around it, its "urban field" (Bryant et al. 1984). The city has grown steadily in recent years (Chapter 4) and this pattern is projected to continue (Pearson 1970:10). One result of its regional centrality is that Brandon exhibits most of the characteristics of larger urban centres, both good and bad, including an urban fringe. The urban fringe, which is typical of the surroundings of other Canadian cities, is now generally accepted as being a region of the city (Bryant 1976; Russwurm 1977a:18; Carter 1981:316). It has been recognized for some time that "the land surrounding our cities is our most valuable body of land still largely unbuilt on" (Russwurm 1977a:9; 1977b:181), resulting in a large body of research on the topic during the last decade. However, as Reeds (1982:2) notes, "in spite of excellent progress to date (by researchers on the fringe), much remains to be accomplished."

The purpose of this chapter on Brandon's urban fringe is not to add significantly to the slowly building "wall of theory" on the fringe but to check how Brandon fits into the established patterns, and thereby to provide more data for urban fringe researchers. This will be done by first discussing the general characteristics of the urban fringe and how the situation at Brandon compares. Second, case studies of different sections of Brandon's fringe will be used to demonstrate how the forces involved in fringe development have a strong impact on even a relatively small, somewhat isolated, urban area. This case study approach is used so that we can describe specific areas in detail, and by so doing demonstrate the problems and character of Brandon's fringe.

## Definition and Characteristics of the Urban Fringe

The urban fringe is the landscape adjacent to urban places of all sizes; it is a changing but permanent part of the regional city extending from the continuous built-up edge of the city outwards into the surrounding municipalities (Russwurm 1977a). Its outer boundary is a topic of some contention and is defined in a variety of ways by different authors (Martin 1975:42). It has been delimited using specific measures such as population density, land use, or land ownership characteristics, but all of these are susceptible to the individual judgements of individual researchers. For large cities, such as Toronto or Montreal, the urban fringe extends at least fifty kilometres, but for smaller cities it is clearly a smaller zone (Russwurm 1977a:13).

As Russwurm indicates, "three notions underlie the general definition" of the urban fringe, and all "are linked directly to the accessibility of the urban fringe land space to our cities" (1977a:13). First, the urban fringe is a spatial transition in terms of land use, population density, and landscapes. Thus parts of the rural municipalities of Cornwallis and Elton adjacent to the city of Brandon (Figure 8:1) have a mixture of land uses, much higher population densities than the surrounding agricultural areas, and landscapes that in places are virtually suburban in form. They are inhabited in some parts by exurban populations engaged, at best, in part-time or hobby farming (Bugg 1982, 1983; Everitt et al. 1985; McCrae 1980).

Second, the land in the urban fringe is undergoing change on a continual basis — it is dynamic (Pryor 1968). Thus the areas around Brandon have shown variation from one period to another. This will be demonstrated later in our case studies, a clear example being the northern fringe which was essentially rural to the mid-1950s, but which is now largely urban in its landscape form.

Third, the fringe is characterized by a mixture of land uses which contrast, often quite markedly, with the more homogeneous zones of consistent land use found elsewhere in the city. Again the northern fringe, which exhibits a great variety of land uses, can be used as an example. The pattern of land uses has changed considerably in recent years; land uses that were extant ten years ago have disappeared and some that were only found close to the commercial core a decade ago are now located close to this edge of the city.

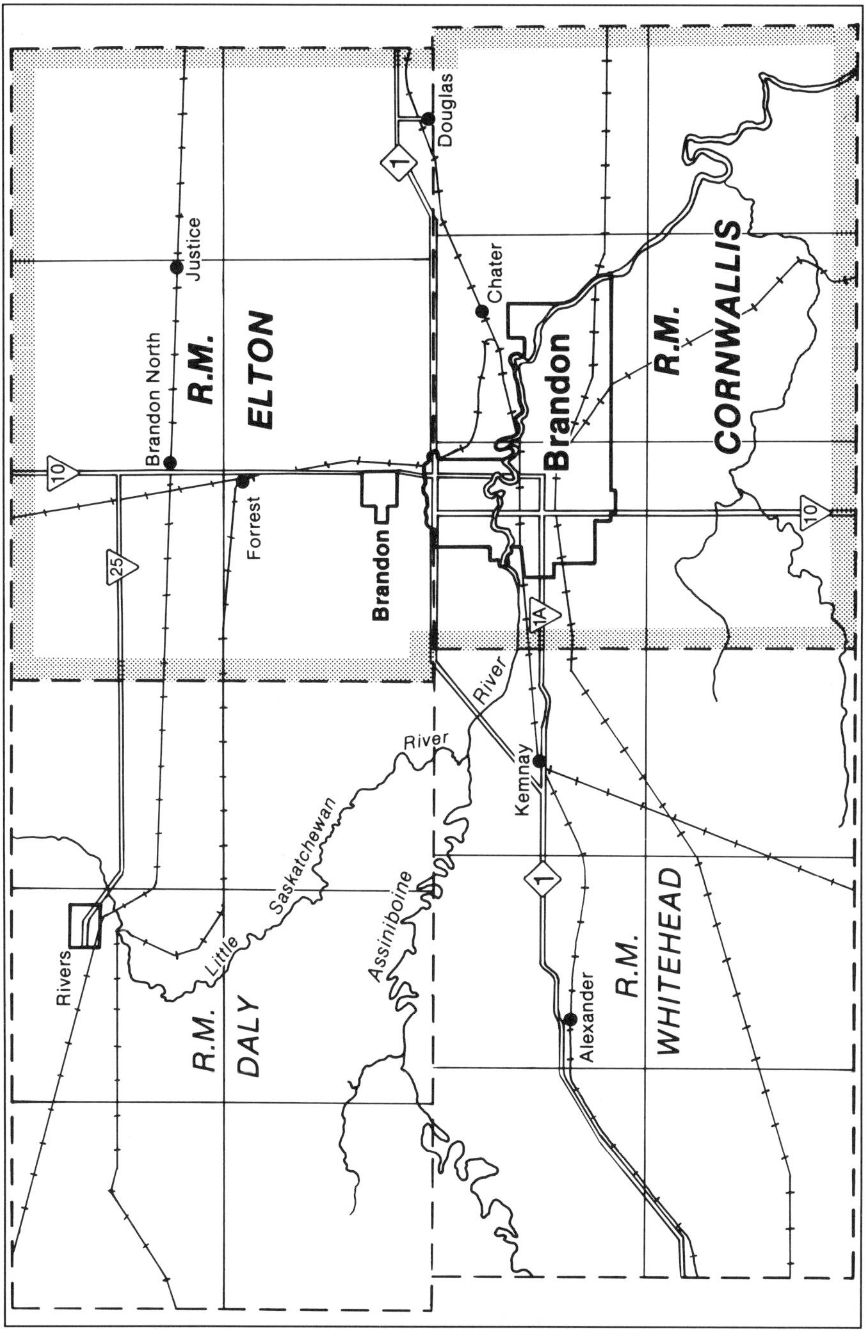


FIGURE 8:1
*The Brandon Area Planning District*

It is clear that the urban fringe is a complex area both for study and for policy making. Therefore, it is not surprising that it is the area in which the greatest conflict between potential land uses takes place (Manning 1980).

## Brandon's Urban Fringe

For the purpose of the present study, Brandon's fringe will be taken to extend about eleven kilometres to the north, east, and west, and about sixteen kilometres to the south. The first boundary roughly coincides with villages which are clearly and closely tied to Brandon's social and economic sphere of influence on a more or less daily basis. The southern boundary is more distant as it takes in parts of a particular physical phenomenon, the Brandon Hills, which for environmental reasons has become both a residential and a recreational extension of the city.

Even the relatively small Brandon fringe has considerable spatial variations resulting from differences in the physical environment, transportation opportunities, socio-economic characteristics, development, zoning regulations, real estate dynamism, and administrative structure. Consequently, it is convenient to consider Brandon's fringe as four distinct areas — the north, the south, the east, and the west. The northern fringe is different from the others as it has been shaped to a greater extent by forces external to the city. These are commonly of a larger, regional nature and are symbolized by the Trans-Canada Highway. The northern fringe is the area where the most significant changes have taken place. The southern and western fringes can be viewed individually but can also be considered as parts of a homogeneous area in which city-related forces have been important in shaping land use change. The eastern fringe is a separate unit, having a unique pattern of development. The four zones will be described in detail, but first, we will provide some introductory remarks aimed at setting the Brandon fringe in its contemporary situation with respect to planning problems and prospects.

### *Planning in the Brandon Fringe*

Although planning is by no means a new concept in our study region, with planning schemes in Brandon, Cornwallis, and Elton dating back to 1948, 1957, and 1972 respectively, unified planning is a more recent development (Chapter 9). The rural municipalities were both members of the advisory Grand Valley Planning District, but joint planning with the city of Brandon began seriously only with the advent of the Brandon Area Planning District (BAPD) in 1980 (Figure 8:1). This group relationship was a recognition of the fact that the three administrative units are integrally entwined in a rural-urban system in which change in one unit can have significant positive and negative effects on the others. The aims of the *Brandon Area Planning District Development Plan* (Province of Manitoba 1984) were to preserve what was viewed as good in the region (e.g., natural resources and valuable agricultural land) while allowing for the growth of other forms of development (nonfarm, particularly residential) outside the city boundaries.

With respect to the fringe developments, a number of principles are worthy of mention. First, rural residential growth was recognized by the Brandon Area Planning District Development Plan (BAPDDP) as a legitimate form of land use, and as an alternative lifestyle to urban residential living. It had been discovered that the people moving into Brandon's fringe areas were doing so for predominantly noneconomic reasons related to the rural character of the fringe. The BAPDDP was sympathetic to these reasons as long as development was "in a well-planned and logical manner and on an economically sound basis" (Province of Manitoba 1984:II-14). Therefore residential building is restricted to areas where it does not conflict with agricultural land use, one result being that specific areas have been designated "rural residential" with a lot size ranging from a minimum of two acres along Grand Valley Road to a maximum of forty acres in a number of locations in Cornwallis.

Second, it was recognized that the "Rural Municipalities of Cornwallis and Elton are primarily oriented toward an agricultural and related resource base." Consequently a major aim of the BAPDDP was "to ensure the protection and logical development of agricultural land uses, to the long-term interest and benefit of the Planning District" (p. II-4). This aim is to be achieved by a variety of policies including zoning, development agreements, and "such programs as are determined from time to time by the Board or Councils to be necessary and in accordance with the provisions of any Part of the Plan" (p. II-11). Such policies should have a major effect on Brandon's fringe by restricting future nonagricultural developments to areas of poorer quality farm land or at least to areas where disruption is controlled and minimized.

Third, the BAPDDP aims to "foster the growth" of existing rural settlement centres such as Chater and Forrest, and fourth, it tries to control "highway and agro-related commercial development" although there has been little of this within the planning district. In addition a series of policies was laid down to cover other forms of land use ranging from aggregate extraction to historical resources, although few such uses have been of areal significance.

*The Northern Fringe*

The northern fringe stretches from the old (pre-1972) city boundary along Braecrest Drive in the south approximately to Forrest in the north, a distance of about eleven kilometres. It runs from the Little Saskatchewan River Valley in the west to about Justice in the east (Figure 8:1). Different levels of intensity of fringe development exist within the area, much of the outer part consisting of relatively level farmland with scattered large lot residential dwellings, that could be termed exurban, associated particularly with Provincial Trunk Highway 10 and the Trans-Canada Highway. In contrast, the inner part consists of a more mixed and intensive set of land uses including the Brandon Municipal Airport and a significant amount of strongly highway-oriented commercial activity. Because the development of the inner fringe is more intensive, and potentially much more of a problem in the overall planning framework of the city, it is our primary concern here.

Fringe development has always been integrally associated with the transportation infrastructure. Although they were initially important, railways have declined in significance since the 1930s as forces stimulating the growth and shaping the urban structure of prairie settlements. Accessibility to roads and highways, particularly paved highways, is now the major factor in the growth process at the edge of a city. This is particularly noticeable in Brandon's northern fringe, which has been greatly affected by the Trans-Canada Highway (Everitt and Stadel 1978), and in which a typical fringe land-use succession pattern has occurred (Barke 1976).

The original route of the Trans-Canada Highway roughly paralleled the CPR tracks from Winnipeg to Brandon. However, it deviated from this line east of Brandon where it skirted north of the Assiniboine floodplain continuing west along the bluffs, following a surveyed township line. It then entered Brandon by running down the northern bluffs of the valley (North Hill) along First Street North crossing the Assiniboine River at one of the two bridging points within the city (Chapter 2). It then continued west through the city towards the Manitoba/Saskatchewan boundary.

As discussed in Chapter 4, the 1882 city boundary was placed around an optimistically large area of land, reflecting the boom-time aspirations of the first city council (Figure 8:2). In the north it followed a section line along what is now Braecrest Drive, a boundary which proved suitable until at least the 1950s. During the 1950s and 1960s, however, an increasing amount of development occurred on the North Hill, spurred on by the completion of the Trans-Canada Highway By-Pass which was placed one mile north of the 1882 boundary. In order to contain and control the development, and to capture its taxable potential, the boundary of the city was moved to a line just north of the new Trans-Canada Highway on 1 January 1972, in spite of strong objections by the rural municipalities of Cornwallis and Elton. An exclave containing McGill Field, the local airport, was incorporated into the city at the same time.

As a result of this political action most of the inner northern fringe was placed administratively within the city of Brandon, bringing the development of the area under the control of one authority rather than three. More recently this area has been included within the Brandon and Area Planning District (Chapter 9) so now most of Brandon's northern fringe, both inner and outer, is under the control of one board (Figure 8:1). The exception is an area to the west which lies in Whitehead Rural Municipality where exurban growth of primarily residential units is occurring with only limited control.

In studying the growth of the inner northern fringe, which in many ways precipitated both the movement of the city boundary and the formation of the planning district, it is useful to look at three time periods: pre-1956 (before the Trans-Canada Highway By-Pass), 1956 to 1972, and post-1972. In each case special reference will be made to the important fringe characteristics of land-use change and land ownership change.

### Pre-1956

During this period, the area north of the 1882 boundary was an agricultural zone, little affected by economic forces from within

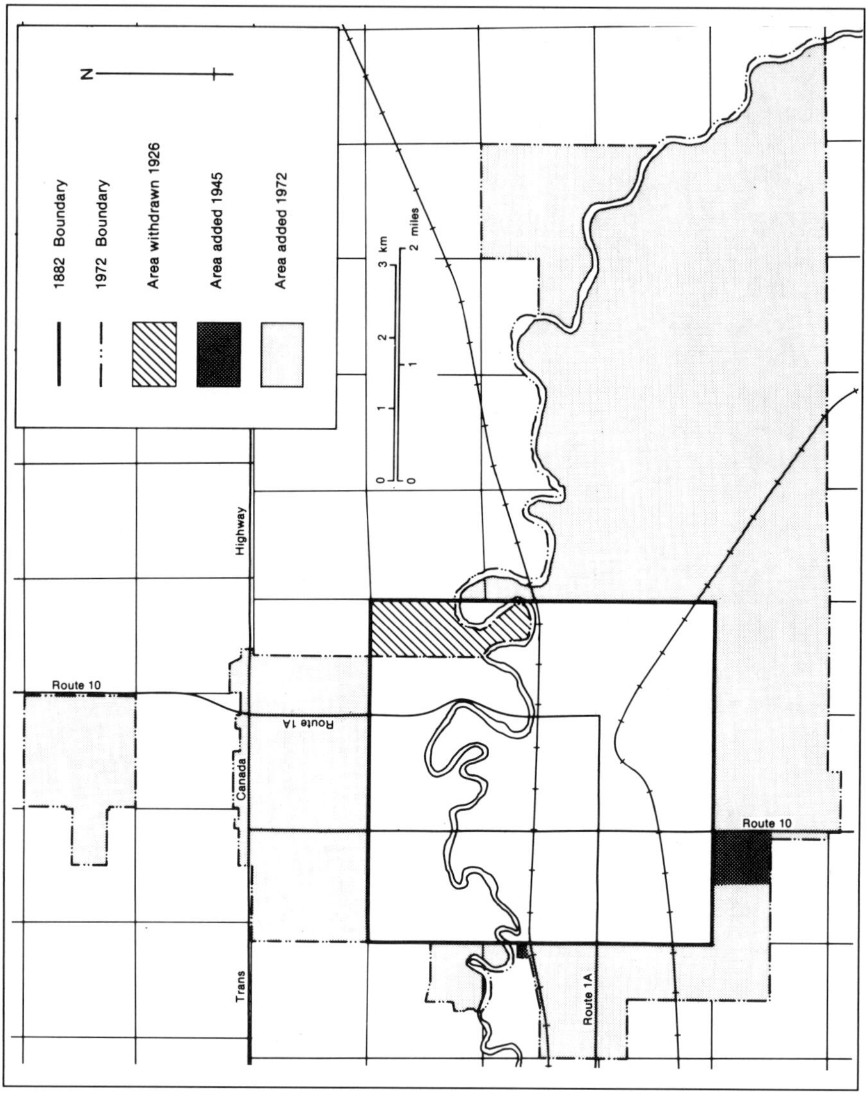


FIGURE 8:2
*Boundary changes for the city of Brandon*

Brandon. Land-use and land-ownership patterns were simple and quite stable, speculative processes were almost nonexistent, the road network was skeletal, and the land was mostly Crown land or was owned and used by local farmers. A noticeable exception to this simple situation developed in a section north of the old Trans-Canada Highway where land was acquired in 1950 to develop a drive-in theatre. Also during this period, a few commercial establishments located on the western side of Highway 1A leading into the city. However, land values generally remained low as the land had not yet become a speculative commodity and it contained few buildings.

1956-1972

In 1956 construction began on the Trans-Canada Highway By-Pass marking the onset of a period of speculation that included increased dynamism in land transfers, rising land values, and land subdivision. Land ownership after 1956 became fragmented, most of the titles being acquired by nonfarmers, although large parts of the land, with the exception of the road and highway ribbons, remained in agricultural use.

By the late 1950s a number of developments had occurred which are worth noting. First, although there was some increase in residential development, the major change was in the form of nascent commercial and industrial land use and the development of gravel pits. Second, some building occurred just outside the city boundary on the west side of First Street North (commercial) and along Braecrest Drive (residential). Third, there was no development east of First Street North which was, and still is, provincially owned land, and there was little land use change along Eighteenth Street North which was still unpaved.

By the early 1960s there had been some major variations in these patterns. The ribbon on the west side of First Street North (Highway 1A) was soon commercially subdivided although the barrier of the Crown land east of the highway remained intact. In addition, commercial development had taken place at the junction of the Trans-Canada Highway and Highway 10, and was beginning to spread west along the Trans-Canada Highway itself. However, much of this commerce was still geared to the Brandon and local area markets. Although Eighteenth Street North had now been paved it still lacked development, but west of the road there was some new residential activity along Braecrest Drive.

By the mid-1960s these changes were increasing in pace. In particular there was more development around the intersection of Eighteenth Street North (Highway 1A) and the Trans-Canada Highway (Highway 1A becomes Highway 10 north of the Trans-Canada Highway), and subdivision had also taken place along Eighteenth Street North itself.

By the late 1960s the acceleration of activity was even more evident. Residential subdivision along Braecrest Drive was continuing although still only to the north, as the federally owned Agricultural Research Station formed an almost complete barrier to the south. Commercial development along Eighteenth Street North was more advanced, and was more "regional" and less "local" in its orientation, with farm equipment and construction-oriented businesses becoming typical. There was more building along the Trans-Canada Highway, with the result that within Section 35 farmland was being encircled by commercial growth on all four sides (Figure 8:3).

In summary, during the period following the construction of the Trans-Canada Highway By-Pass, the pace of land transfers increased significantly, particularly along the highway ribbons. Between the late 1950s and early 1970s, the most dynamic area in terms of land transfers, incipient speculation, and land-use change was at the prime intersection of the Trans-Canada Highway and Highway 10/ Highway 1A; for example, between 1957 and 1970, the two parcels on the north side of the intersection were subject to a flurry of nine land transfers involving change of ownership from a Brandon road builder, to a Winnipeg-based oil company, to another Brandon builder, and finally to a series of motel owners. The sworn value, which gives a rough indication of the land value, of the two parcels increased from $7,460 in 1957 to $90,000 in 1959, $125,000 in 1963, $220,000 in 1967, and $275,000 in 1971. Obviously the economic potential of this area around the city was finally being recognized.

### Post-1972

The northward shift of Brandon's boundary to include the whole of Sections 34 and 35 and part of 36, and the ribbons along the Trans-Canada Highway By-Pass, brought almost the whole inner fringe within the legal boundaries of the city (Figure 8:2). This

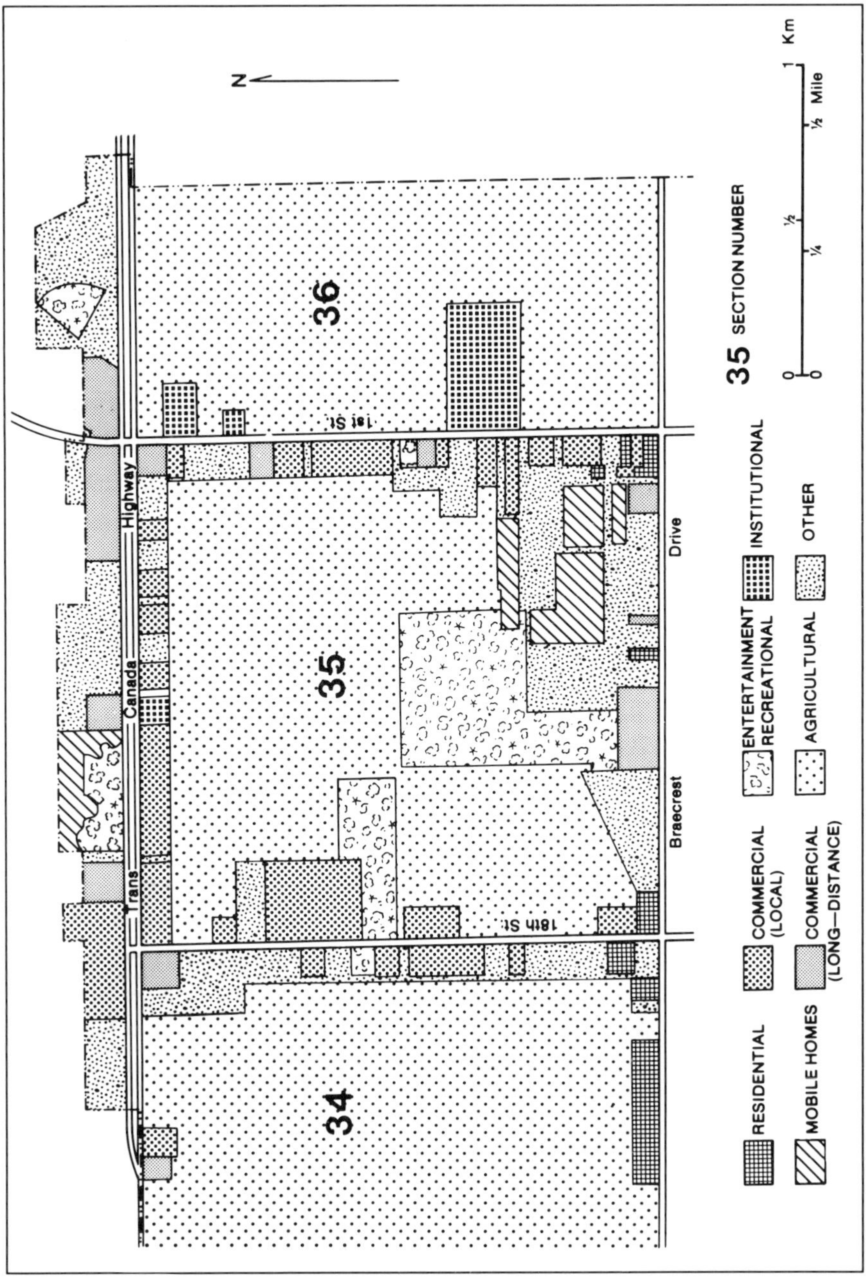


FIGURE 8:3
*Land use on Brandon's northern fringe: 1979*

development, together with new zoning bylaws introduced by the adjacent rural municipalties of Cornwallis and Elton to prevent urban sprawl, stimulated residential and commercial building. Furthermore, the new BAPDDP clearly favours a policy of "residential infilling" on the North Hill, a process that is slowly occurring. The most significant developments in the northern inner fringe since 1972, along with the contemporary situations, can be stated as follows: (1) West of Eighteenth Street North, the north side of Braecrest Drive has been almost completely filled with housing, the "gaps" reflecting the presence of ravines, a hydro line, and one or two empty lots. Here a new residential area is located on the scenic, partially wooded, land in the southeast corner of Section 34 (Figure 8:3). This land, a parcel of about forty-five acres with a sworn value of $108,000 in October 1978, was jointly developed by two local farmers, the original owners of the land, and by an investment company. Since 1978, the Braecrest Estates (or Hamilton Heights) area, with minimum one-acre lots, has become one of the most exclusive residential areas of the city. Land ownership records confirm the high-class status of the area; by early 1979 individual residential lots were being sold at $20,000-$23,500 depending on size and site characteristics; by 1986 many lots had been built on, some speculatively, but empty lots were selling for as much as $40,000. Nearby, a second upper income suburban-type development, undoubtedly encouraged by the success of Hamilton Heights, is now under way on the only non-government owned land south of Braecrest Drive at its intersection with Eighteenth Street North. (2) For the first time significant building has occurred east of Eighteenth Street, following subdivision of the agricultural land in the area. However, the land use is still not very intensive (e.g., machinery storage) and several empty lots still exist. West of Eighteenth Street the commercial strip is more continuous, although relatively less activity has taken place since 1972. (3) Mobile home development has taken place in the southeastern part of Section 35; in particular a 117-lot mobile subdivision, Highland Park, complete with underground wiring, sewers and city water has been established. The owner is hoping eventually to redevelop the area as one of conventional housing. In addition to Highland Park, there is also the older White Swan Mobile Home Park, with 123 lots, which is owned by the same company. (4) East of Eighteenth Street, immediately north of the 1882 city boundary, along Braecrest Drive, a highly diversified pattern of lot or parcel sizes, land ownership, frequency of land transfers, and sworn values has occurred. A large parcel includes the North Hill Motel owned by a locally based company, west of the motel lies an abandoned sand and gravel pit owned by a local contractor, and to the east the land is subdivided into small residential and commercial lots. (5) The interior of the northern half of Section 35 was acquired in 1975 by the Manitoba Housing and Renewal Corporation as a land bank. It now figures prominently in the city's long-term development plans for the northern part of Brandon. (6) Nearly all of the west side of Highway 1A (First Street North) has been built up, the functions changing from time to time, but the land to the east of the highway remains under government ownership. (7) There has been infilling along the south side of the Trans-Canada Highway, including a recreational vehicle ownership, a mobile home sales area, and a large Ukrainian restaurant. On the north side, where there is only limited development, land use is predominantly oriented to the potential of a "transient market" generated by the highway (Plate 8:1). The differential extension of urban services (water, etc.) north to the Trans-Canada Highway contributed to this pattern of change. Only since the inclusion of the Trans-Canada Highway into the city of Brandon has the commercial investment potential of the zone along the highway, as well as that along First Street North and Eighteenth Street North, been fully realized.

The building along the Trans-Canada Highway is different from that in the rest of the northern fringe in two ways. First, most of the development is geared to either long-distance transit traffic, or to activities which have a great demand for space and/or need easy accessibility. Many of these functions are also able to take advantage of the "free advertising" potential afforded by the view from the highway, and also of the convenience of access, important to mobile home and travel trailer dealerships. Second, the ribbons along the highway are the locations in which nonlocal ownership plays its most significant role, with national as well as international organizations dominating the economic landscape.

## Conclusions about the Northern Fringe

Four major concluding points can be made about the inner northern fringe: (1) The shaping of the pattern of land ownership, land transfers, land use, and sworn values has been determined by a

PLATE 8:1

*This view, from the northeast, shows part of Brandon's northern fringe near the intersection of the Trans-Canada Highway and Eighteenth Street North (Highway 10). A good cross section of fringe activities is shown with farms and farmland still dominating the area. In the foreground, north of the TCH, are located a mobile home park and a motel/restaurant/recreation complex. The display lot of an agricultural equipment sales firm can be seen at the western edge of the area. On the southwest corner of the road junction are a gas station (with a large Canadian flag) and a rendering plant. South of the TCH are a mobile home sales lot and a trucking firm. Highway 10 is flanked by an agricultural equipment sales firm, a truck sales firm and a golf driving range on the east, with woodyards and other regionally oriented commercial enterprises on the west. Expensive treed residential estates can be seen in the background (top left of the photograph). (John Everitt, 1987)*

set of factors and forces, the most significant of which are: (a) the specific site characteristics of the area; (b) the building of the Trans-Canada Highway By-Pass and its arterial highway links with the centre of Brandon; (c) the northern boundary of the city and its movement northwards in 1972; (d) planning priorities and municipal bylaws of both the city and the adjacent rural municipalities; and (e) decisions by the principal actors in the land transfer process, which do not always reflect the fundamental economic processes involved in land development (Everitt and Stadel 1980). (2) The relative importance of these factors and processes has led to the growth of particular commercial, residential, and recreational sub-areas, each one characterized by a specific pattern of development. (3) Much of the construction of the economic landscape was carried out by local residents and companies. The most significant actors in the land transfer process were often farmers who became real estate agents using their farmlands as a speculative commodity. (4) The efforts of the planners in this area have not showed major success. At one stage, the province of Manitoba attempted to introduce a second bypass to circumvent the present Trans-Canada Highway business strip (*Brandon Sun* 25/01/1984). This was an unpopular cause within the city and was eventually shelved.

*The Southern Fringe*

The southern fringe has some similarities to that in the north, but it also has a number of significant differences. The similarities include a complex mix of land uses ranging from farm equipment sales to mobile home parks; an orientation towards Highway 10, which also passes through the northern fringe; an incomplete pattern of land coverage; and a similar era of change since the mid-1960s (Vestby 1979). The major differences are tied to the raisons d'être of this area which is not so much a product of highway development as is the area to the north. Here, the major forces are commercial developments and the existence of the Brandon Hills as an attractive physical environment. An additional difference is one of scale: the northern fringe is centred on three arterial routes whereas the southern commercial strip is based upon the area along Eighteenth Street (Highway 10) and its junction with Richmond Avenue (Plate 7:4).

The commercial developments that have occurred are varied but many owe their success to the opening of the Brandon Shoppers' Mall on the extreme edge of the built-up area in 1972 (Chapter 7). Since then the mall has expanded, but perhaps more importantly several symbiotic commercial and residential developments have located close by, including a number of fast-food restaurants and a Canadian Tire store. A secondary mall has been mooted recently. These commercial outlets, together with the removal of the oil refinery at the junction of Richmond Avenue and Eighteenth Street (Plate 6:1), spurred on residential development, the whole pattern being reinforced by the opening of a regional high school south of the junction of First Street and Richmond Avenue. This array of development has had a multiplier effect resulting in most of the south end of the city being at least partially built up, mostly with middle-class housing. The mall acted as a "growth pole," so that much of the area was rapidly filled up by continuous urban land use, resulting in the fringe spreading to include the area south of the Brandon Cemetery lying within the city limits, as well as an area of exurban housing further south.

Few commercial fringe developments exist outside the city limits with only a drive-in movie theatre (recently closed), an antique store, and a few other such services dispersed along Highway 10. This area is still largely an agricultural zone with class 2 and 3 soils, with a few scattered exurban dwellings on large lots which become more frequent as the Brandon Hills (class 4 and 5 soils) are approached. The wooded hills, a recessional moraine, are of marginal value agriculturally but of considerable value for residential dwellings.

Although this southern fringe is located within the two rural municipalities of Cornwallis and Oakland, discussion, based on a research project completed by Sikora (1980), will be limited to the area that lies within Cornwallis as it contains the main activities of the fringe and is most affected by the proximity of Brandon. Cornwallis, which has always been affected by Brandon, has tried strenuously in recent years to retain its agricultural orientation as a counter to the fringe developments that were changing its character, particularly since the 1972 boundary change which took away a large part of its territory. One result was that in 1975 Cornwallis adopted a development plan with a revised version in 1977. Zoning now exists in Cornwallis to "provide for the logical development and protection of general agricultural and related uses [and to] maintain the rural and open space character of the municipality" (Province of Manitoba 1975:2). Cornwallis's membership in the BAPD enables it to fulfill these goals more easily and to avoid the worst aspects of fringe development which are clearly recognized and hopefully protected against.

However, despite these efforts, the rural municipality contains a high proportion of non-farm residences, a figure which may be as high as 60-70 percent in the area immediately south of the city boundary. The migration of people into the fringe — 73 percent were ex-Brandonites in Sikora's study — has not been curtailed by the 1975 zoning restrictions, although they may have diverted migrants from the best agricultural land.

The reasons for the out-migration from Brandon are typical of those cited by other fringe researchers, such as a desire for more space, a love of the outdoors and/or a wish to take up farming or hobby farming. It is interesting that even in a small city such as Brandon people may wish to "get away from city life" and live in the country. Surprisingly, obvious economic factors such as lower taxes were rarely cited as reasons for moving.

Similarly, the lifestyles of the fringe residents are typical for fringe dwellers elsewhere. A large proportion of the people (85 percent) were at least partly engaged in agricultural activities, but most of these made only a small (under 50 percent) proportion of their gross income from farming, and only 10 percent were completely dependent upon agriculture for their livelihood. In addition, 70 percent of the heads of households commuted to Brandon on a daily basis and the city was the principal area for schooling, entertainment, shopping, and many other social and economic activities. The sample group was also found to be relatively wealthy compared to the urban average, indicating that it is the middle class that most commonly moves to the fringe.

This fringe area was not only a convenient place to live, it also provided a satisfying environment, although the responses also indicated a desire for an urban way of life. The major advantages perceived by the study group were: not far from Brandon; offers more space; escape from city life; natural beauty of the site; allows pursuit of hobbies and/or sports; and access to good roads. As before, the economic advantages (16 percent) were outweighed by the social-aesthetic amenities (84 percent).

Disadvantages were also noted, and once again were those common to fringe dwellers elsewhere, such as problems with "snowed roads" and poor "municipal road maintenance." Most of these disadvantages reflected the desire for a high level of mobility and again indicate the respondents' intensive interaction with Brandon.

Sikora (1980) also attempted to profile the "typical fringe dweller" and to compare this "average" with that in other studies and with city of Brandon data. The fringe dwellers in this area correlate quite well with those found elsewhere but differ from the "average" Brandon dweller. Thus 95 percent of the respondents were married, the heads of households averaged about 46 years of age, and there was a completed family size of nearly four people. Most were urban workers with a fairly high family income compared to the average within the city, supporting the findings elsewhere that higher income classes tend to be overrepresented in the fringe (Russwurm 1977b:259). In addition the fringe dweller had a clear bias towards a more rural, bucolic way of life, had backed up this desire by overt action (the move to the fringe), and being well satisfied with the results was unlikely to move back to the city.

Empirical analysis of the fringe south of Brandon revealed that the process of rural transformation is clearly present. Although the rural municipality of Cornwallis has declared most of the area to be an agriculturally oriented community and has zoned it as such, non-farm fringe development continues, for as Hardwick (1972) has demonstrated, zoning by itself cannot guarantee the continuation of farming.

### *The Western Fringe*

The western fringe is defined as an area that extends west of the city boundary approximately as far as Kemnay, particularly the areas along Highway 1A and Provincial Road 459 — the "Grand Valley Road" (Figure 8:1). However, it should be recognized that Brandon's influence has led to fringe development outside this area, for instance along the Trans-Canada Highway and even west of Kemnay, but, as these areas do not add anything substantial to the discussion, they may be neglected without loss of precision. This discussion is based, in part, upon a research paper by Vestby (1980), which concentrates on parts of the rural municipalities of Cornwallis and Whitehead, the former being in the BAPD but the latter lying outside it to the west. The area is one of mixed agricultural potential. South of Highway 1A and within the Assiniboine Valley the soils are mostly of classes 2 and 3; these are generally farmed or designated as Rural Conservation Recreation Areas (within Cornwallis). In contrast, soils on the northern bluffs of the valley are mapped as class 4 or 5.

The western fringe differs from those to the north and south in that its commercial component is less well defined and less significant. Highway 1A, part of the original Trans-Canada Highway, lost much of its prominence in the late 1950s when the bypass was completed to avoid both the city and a low railway bridge at

Kemnay. As there had been little fringe development around Brandon prior to 1958, this region remained a commercial backwater until the westward suburban and exurban push began in the 1970s, continuing into the 1980s (Chapter 4). Recently development has increased and, although it is still not the preferred area for most forms of commercial growth, it has proved successful for high quality motels and some car dealerships. The western fringe has been particularly important for residential building as the two major roads give excellent access to the city resulting in a 60 percent increase in the number of houses between 1976 and 1980.

In many ways the magnetism of the western fringe is similar to that of the area south of Brandon. The major reasons given for moving to this area were to get more space, to take advantage of the outdoors, to take up (hobby) farming, to avoid city taxes, and to avoid city life; 85 percent of the respondents had moved from Brandon. The move was also seen as a good investment, but, once again, non-economic factors were more important than economic concerns when moving.

A smaller proportion of the study group, approximately 30 percent, derived a part of its income from farming activities and only 6 percent was completely dependent upon agriculture, compared with 85 percent and 10 percent respectively in the south. This difference was also reflected in the size of holdings; in the west 20 percent owned under one acre and 60 percent ten acres or less, whereas in the south the comparable figures were 0 percent and 30 percent. This variation is related to the fact that the southern area was wholly within Cornwallis whereas the western fringe also includes part of Whitehead, which is not in the BAPD and in which zoning regulations are less restrictive. In addition the dwelling units were larger in the western fringe, again probably reflecting a larger proportion of middle-class exurbanites who had left the city to build a large unit in the countryside. Also, commuting was more characteristic of the western fringe and the non-working family members were more mobile (to do shopping, to attend schools, etc.). There were no major differences in income between the two areas, but the western fringe dwellers were, on average, better educated, and included a smaller proportion of farmers than was the case in the southern area. Once again these fringe dwellers were more affluent than their urban counterparts, although there were no significant variations from the people in the other fringe areas.

The western fringe was a satisfying, as well as a convenient, environment in which to live as it was rural and yet offered easy access to urban life. The major advantages cited were almost identical to those found by Sikora (1980) in the southern fringe although, perhaps understandably, the people in the west did not value the "natural beauty" of the sites as highly. The disadvantages were also similar, with the low quality of municipal services and distance causing most problems, although 50 percent found no disadvantages to the location. Jackson (1973:201) recognizes these problems as typical of fringe areas when he states, "Many essential services such as water mains, sewers and bus services become prohibitively costly when the population is widely dispersed. . . . The new residents, though liking their spacious, quiet, and clean surroundings, are often dissatisfied by these service deficiencies. Most want to live closer to the City, and they find that the fringes are by no means ideal areas for raising children." However, it appears that in the minds of the western fringe residents, the advantages of exurban living far outweigh any disadvantages.

Vestby, like Sikora, prepared a profile of the typical (western) fringe dweller, the results indicating great similarities between the two areas. Clearly people have personal reasons for choosing one area over the other, but there are indications that a move to the south is more costly than one to the west, perhaps enticing more affluent migrants to the area near the Brandon Hills. But in large part these two groups of fringe dwellers can be considered as similar.

### *The Eastern Fringe*

The urban fringe area to the east is quite different from those to the north, west and south. These differences result from the transportation patterns, the political history of the region, the physical background of the zone, and the existence in this area of the Canadian Forces Base (CFB) at Shilo. Two major east-west roads bisect this fringe — the Trans-Canada Highway and Provincial Road 457.

Politically much of this eastern area has been within the city since 1972 and the railway yards, extraction areas, sewage lagoons and industrial zones are typical fringe uses. When the city was proposing to extend its boundaries in the late 1960s it hoped that the extension would include most of Cornwallis in this direction.

However, this proposal was rejected and much of the region, and most of Provincial Road 457, remained within municipal jurisdiction. Discussion of this eastern fringe will be centred on the area lying within Cornwallis as it is here that the greatest residential impact is found, with a series of rural residential zones of two, five and forty-acre lots. In addition it contains the general development areas of Chater, Douglas and Spruce Woods (the settlement at the gates of CFB Shilo), although these will not be discussed in detail.

Physically the eastern fringe is made up of three zones. The first is a ribbon along the Assiniboine River Valley (with class 2 and 3 soils), much of which is designated in the *Development Plan* as rural conservation recreation area. The second is an area of good agricultural land (also with class 2 and 3 soils) lying to the north of the CPR mainline and containing some residential dwellings. The third is a large area of class 5 soils interspersed with class 4, 6 and 7 soils. In this area agriculture is severely limited and fringe developments have occurred. Exceptions to this general pattern are some minor developments close to Brandon, including the new Cornwallis municipal office, and a mobile home park located on an area of class 5 soil northwest of Chater.

Within this eastern zone most development has taken place along Provincial Road 457 and Provincial Road 340, which connects CFB Shilo with Douglas at the Trans-Canada Highway, the developments being related to these routeways closely enough to be termed corridor developments (Lafleur 1979). As the area is mostly one of poor soils it traditionally supported large farms and large cattle operations. Another characteristic differentiating this area from other fringe areas is the existence of CFB Shilo which operates as a secondary growth pole for both commercial and residential development. Besides the population of the camp itself, which has spurred commercial development along Provincial Road 340 and in the Spruce Woods community, many local non-farm residents work in Shilo and make use of its recreational facilities. There are, then, strong links between CFB Shilo and the surrounding areas which are reflected in the use of services and in the commuter-like transportation patterns to and from the camp.

Most of the non-farm developments in this area, as elsewhere in the areas adjacent to Brandon, are recent. A 1973 municipality of Cornwallis land use map shows only seven non-farm residences (excluding those in Chater) whereas today there are several hundred residents. The number of farmers has dropped over the same period as farm sizes have grown and/or subdivision of the land for residential uses has occurred. Currently there are seven residential areas in eastern Cornwallis ranging in size from the Sprucewoods General Development Area, through an area of two-acre lots, to one which is zoned RR 40 (forty-acre lot minimum size). A development which typifies this fringe is Cottonwood Acres which began in the mid-1970s when about three hundred acres of land were bought by Brandon real estate developers from two local farmers and subdivided into fifty-four lots, a subdivision which was supported by Cornwallis. The rural municipality regarded the division of the land into five-acre lots for hobby farms as an aid to controlling development and preventing piecemeal building throughout the municipality (Hollender 1979). It also had the advantage of taking place on marginal land generating tax income for the rural municipality. It is well connected to Brandon by a good road system so Cottonwood's residents can make good use of the city. Interestingly, this and neighbouring developments make use of Douglas School (to Grade Six) and Elton Collegiate in Forrest (to Grade Twelve) and therefore do not depend upon Brandon for school facilities. It is not clear whether this has either a positive or a negative effect on potential residents when compared to the zones to the west and south of the city.

The study by Lafleur (1979) revealed that the residential dwellers who have moved to this area as an alternative to Brandon have done so for many of the same reasons as those already listed for the western and southern fringes. The non-monetary attractions of rural life were important but the financial advantages of lower taxes were also noted. The disadvantages of long distance (fifteen to twenty kilometres) commuting and poorer services were documented but obviously were not of great significance for those who made the move. Once again the perceived attractive lifestyle would not be possible without the presence of Brandon which remains important for shopping and workplace locations. The influence of the city has been so strong that neither Chater nor Douglas serves as a central place except for a few low-order functions (churches, recreational facilities and small stores). Only the Sprucewoods Community has several services reflecting its proximity to Shilo, to the rural residential developments in this area, and to the recreational potential of the region.

## Conclusion

In his landmark work, *The Surroundings of Our Cities*, Russwurm (1977a) identified eight "urban fringe problems" that are common throughout Canada. These were conflicting land-use activities, land conversion difficulties, impact on agricultural land, impact on the natural environment, services versus taxation (who benefits, who pays?), impact on surrounding settlements, social issues, and governmental planning and administrative difficulties. These problems were substantially the same as those identifed by the *Brandon and Area Planning District Development Plan* (Province of Manitoba 1984) as having characterized the Brandon urban fringe for many years.

Many of these problems can be still seen in the local area either actively (for example, services versus taxation) or perhaps, as a result of the *Development Plan*, in a relic form. Unfortunately the BAPD may be too small to solve the long-term problems. In particular the fact that Whitehead to the west did not join could prove to be a severe drawback, and even Oakland and the southern part of Cornwallis to the south, and North Cypress to the east, may be under pressure. It has been demonstrated elsewhere that conditions such as these found in the Brandon area simply cause the problems to leapfrog to uncontrolled jurisdictions; there is some evidence that this is happening. It is hoped that the heightened awareness of the need for planning, in both rural and urban areas, that now appears to exist will help to prevent, or at least short-circuit, the major problems before they get out of control.

## References

Barke, M. 1976. Land-use succession: a factor in fringe-belt modification. *Area* 8(4):303-6.

*Brandon Sun*, 25 January 1984.

Bryant, C.R. 1976. *Farm-Generated Determinants of Land Use Changes in the Rural-Urban Fringe in Canada 1961-1975.* Ottawa: Environment Canada, Lands Directorate.

Bryant, C.R., L.H. Russwurm and S.Y. Wong. 1984. Agriculture in the Canadian urban field. In M.F. Bunce and M.J. Troughton, eds., *The Pressures of Change in Rural Canada.* Toronto: York University, Department of Geography Monograph 14.

Bugg, R.G. 1982. Brandon's Fringe and Shadow. Student paper, Department of Geography, Brandon University.

Bugg, R.G. 1983. A Typology of Farming in the Rural Municipality of Elton. B.A. thesis, Department of Geography, Brandon University.

Carter, H. 1981. The rural urban fringe. In H. Carter, ed., *The Study of Urban Geography.* London: Edward Arnold.

Everitt, J.C. and C. Stadel. 1978. The impact of the Trans-Canada Highway on the urban fringe of Brandon, Manitoba. Annual Meeting of the Canadian Association of Geographers, London, May 1978.

Everitt, J.C. and C. Stadel. 1980. A study of power and politics in the urban fringe of Brandon. In A.A. Seaborne, ed., *Some Current Research on the Canadian Plains.* Regina: Department of Geography, Regina Geographical Studies 3.

Everitt, J.C., C. Stadel and R.G. Bugg. 1985. A farm typology of agriculture in the vicinity of Brandon, Manitoba. Annual Meeting of the Prairie Division of the Canadian Association of Geographers, Kenosee Park, Moose Mountain, Saskatchewan, September 1985.

Hardwick, W.G. 1972. The Georgia Strait urban region. In J.L. Robinson, ed., *British Columbia, Studies in Canadian Geography.* Toronto: University of Toronto Press.

Hollender, B. 1979. Urbanisation of the Southwestern Manitoba Landscape: A Study of Cottonwood Acres. Student paper, Department of Geography, Brandon University.

Jackson, J.W. 1973. *The Canadian City: Space, Form, Quality.* Toronto: McGraw Hill-Ryerson.

Lafleur, P. 1979. A Study of The Brandon-Shilo Urban Corridor. Student paper, Department of Geography, Brandon University.

Lewis, G.J. and D.J. Maund. 1976. The urbanisation of the countryside: a framework for analysis. *Geografiska Annaler* 58B:17-27.

Manning, E.W. 1980. *Issues in Canadian Land Use.* Ottawa: Environment Canada, Lands Directorate, Working Paper 9.

McCrae, J.D. 1980. *The Influence of Exurbanite Settlement in Rural Areas: A Review of the Canadian Literature.* Ottawa: Environment Canada, Lands Directorate, Working Paper 3.

Martin, L.R.G. 1975. *A Comparative Urban Fringe Study Methodology.* Ottawa: Environment Canada, Lands Directorate, Occasional Paper 6.

Pearson, N. 1970. *The Future of the Middle-Size City, Satellite, Ghost Town, or Human Community.* Guelph: University of Guelph, Centre for Resources Development, Publication 3.

Province of Manitoba. 1975. *Rural Municipality of Cornwallis Development Plan Report.* Brandon: Province of Manitoba, Municipal Planning Branch.

Province of Manitoba. 1984. *Brandon and Area Planning District Development Plan.* Brandon: Province of Manitoba, Municipal Planning Branch.

Pryor, R.J. 1968. A sampling frame for the rural-urban fringe. *The Professional Geographer* 20(4):257-61.

Reeds, L.G. 1982. Rural-Urban fringe studies: a critical review. In K.B. Beesley and C. Cocklin, eds., *Perspectives on the Rural-Urban Fringe.* Guelph: University of Guelph, Occasional Papers in Geography 2.

Russwurm, L.H. 1977a. *The Surroundings of Our Cities: Problems and Planning Implications of Urban Fringe Landscapes.* Ottawa: Community Planning Press.

Russworm, L.H. 1977b. The urban fringe as a regional environment. In L.H. Russwurm, R.E. Preston and L.R.G. Martin, eds., *Essays on Canadian Urban Process and Forms.* Waterloo: University of Waterloo, Department of Geography, Publication Series 10.

Sikora, H. 1980. The Southern Fringe of Brandon. Student paper, Department of Geography, Brandon University.

Vestby, B. 1979. Rural-Urban Land Conversion in Brandon, Manitoba 1964-1979. Student paper, Department of Geography, Brandon University.

Vestby, B. 1980. Profile of a Typical Exurbanite in Brandon's Western Fringe. Student paper, Department of Geography, Brandon University.

# NINE
# Urban Planning in the City of Brandon

*Ron Tompkins*

The field of urban planning is primarily concerned with managing the growth and development processes of towns and cities. This ultimately involves a process of managing the physical arrangement of all aspects of urban development — including various residential, commercial, industrial and institutional buildings; transportation systems such as roads and railways; as well as various open spaces — in a manner which provides for the mutual compatibility of all development and the general well-being of the population.

In order to understand the evolution of urban planning in Brandon, it is important to have a basic understanding of the general evolution of the urban planning tradition and of the history of Brandon's growth and development. The first section of this chapter contains a synopsis of the evolution of the urban planning tradition. The second section contains a brief history of the growth and development of Brandon, as seen from a planning perspective. The evolution of urban planning in Brandon since 1948 and a brief overview of the current planning program are presented in the third section, and the fourth section provides some glimpses of the future.

## Evolution of the Urban Planning Tradition

To a large extent, contemporary urban planning can be viewed as the culmination of two interrelated traditions — urban design and zoning. The urban design tradition can be traced back to the origins of Western civilization. During the ancient Greek and Roman periods, the gridiron plan, which featured standardized rectangular blocks and a street system intersecting at right angles, was developed and widely used, particularly in the development of ancient colonial settlements (Mumford 1961:191-200). The period up to the twentieth century was characterized by rapid settlement and expansion in North America. During this era, the gridiron plan was widely used as a pattern of urban design, as it could be quickly surveyed, provided standardized units of land, and thereby facilitated rapid settlement and occupation (Mumford 1961:421-26). Therefore it is not surprising to find that major portions of most of the communities in North America are characterized by the rectangular gridiron plan. The gridiron plan has been a dominant design feature in Brandon, particularly in the older portion of the city bounded by First Street, Richmond Avenue, Twenty-Sixth Street and Pacific Avenue.

The period of rapid urban growth and the development of automobile technology during the twentieth century have revealed a number of the gridiron's shortcomings. Its rigid geometry has not been readily adaptable to a variety of terrain conditions, such as sloping hillsides and irregular river banks. Its uniform street system has also been inadequate in terms of its capacity to accommodate ever-increasing volumes of vehicular traffic. From an aesthetic point of view, the gridiron has also been criticized as being too monotonous, with its visually endless streets, flanked by uniform rows of buildings (Mumford 1961:421-26).

After World War II, many of the new designs for urban areas deviated from the traditional gridiron pattern. The new designs were more compatible with terrain conditions, and featured a hierarchy of curvilinear street systems, from low-speed residential streets to high-speed arterials. These new designs provided for more attractive developments and a more rational transportation system. The new urban design concepts became apparent in those areas of Brandon which were developed during the postwar era, such as Valleyview, Linden Lanes and Green Acres, as well as in more recently developed areas such as Riverheights, Westview and Parkdale Heights (Figure 4:4).

In contrast to the long evolutionary period of the urban design tradition, the zoning tradition is essentially a twentieth-century phenomenon. The roots of the zoning tradition can be traced back to the period of industrialization in the second half of the nineteenth century. The cities of this era were typified by crowded tenement housing juxtaposed with factories. The "smokestack industries" of the time produced excessive amounts of noise, smoke and filth which created unpleasant conditions for nearby residents. The high-density housing failed to provide basic amenities such as sanitation, sunlight, ventilation, privacy and recreation. Inadequate water supply and sewage disposal practices contributed to outbreaks of highly contagious diseases such as typhoid and cholera. All of these factors contributed to urban environments which were not only unattractive, but genuinely unhealthy and intolerable (Mumford 1961:446-78).

These conditions produced an increasing public awareness that cities were becoming intolerable human settlements, and the realization that the processes of urban growth and development must not be allowed to continue in such an unrestrained fashion. In the early twentieth century, urban zoning was introduced in Britain and the United States as a means of organizing and regulating the urban development process. The development of zoning techniques proceeded rather quickly and much of the Canadian tradition was borrowed from techniques developed in these countries (Gerecke 1977:150-51). Many of the early zoning regulations were intended to segregate various types of development, particularly residential development and heavy industry, and also to regulate the density of residential development. Zoning found relatively quick acceptance as a means of managing urban growth processes (Goodman and Freund 1968:21-24), and has been continuously modified during the past seventy years as an instrument for managing urban development.

Prior to this century, urban planning was not recognized as a separate field of endeavour. In most cities, planning was a very informal process, and was undertaken by surveyors, who laid out new urban areas according to the gridiron plan (Gerecke 1977:150). Within the geometric framework created by the surveyors, other planning decisions were made by developers, architects, engineers, builders and other individuals and groups directly involved in the urban development process. Often, such decisions were made in isolation, in the interests of the individuals or groups directly involved, and not always in the best interests of the community. Until recently, the process of urban planning was not highly formalized, and consisted of an agglomeration of piecemeal decisions made by various individuals and groups.

The formal urban planning process did not emerge as a field of endeavour distinct from surveying, engineering and architecture until the twentieth century. This was largely due to the fact that city governments had little inclination or authority to influence the process of urban development. In response to the appalling conditions which emerged in the industrial cities during the nineteenth century, several countries, notably Britain and the United States, introduced enabling legislation which provided city governments with the authority to take a more active role in the process of urban development. The emergence and evolution of the formal urban planning process, and urban planning as a distinct profession, largely coincided with the evolution of the zoning tradition during this century.

In the early part of this century, urban planning followed two interrelated courses — first, the preparation and administration of zoning regulations to control the location and density of new development, and second, the refinement of the urban design tradition, through the concept of the "master plan." As the century progressed, the scope of urban planning increased and it became more widely applied and accepted. The economic and social problems of the Great Depression resulted in an increased focus on economic and social issues in the urban planning process (Goodman and Freund 1968:26-28). The unprecedented economic prosperity of the postwar era unleashed a surge of automobile ownership and housing construction, which in tandem created the prolific growth of the suburbs in virtually all North American cities. Associated with this growth were massive public expenditures required for urban services and infrastructure, such as public transit, schools, roads, sewers, water supply and electrical power. Planning became more important as a means of coordinating this multitude of factors, as well as organizing the pattern of new development.

In recent years, urban development pressures have expanded beyond traditional urban boundaries. The proliferation of automobiles has enabled many urbanites to establish homes in the countryside and to commute to urban areas (Chapter 8). In addition, the rural area has increasingly been utilized to fulfill other urban needs, such as recreational areas, transportation (airports), water supply reservoirs and sewage disposal facilities (Russwurm 1977:39-51). This trend of urban dispersal has contributed to some significant rural problems, including major losses of prime agricultural land and operational constraints for agricultural operations. The growing environmental movement of the past twenty-five years has focussed increasing attention on the problems of the rural environment and rural land-use issues. In order to respond to these emerging rural problems, urban planning expanded beyond its traditional urban boundaries to become a regional or district planning process. Locally, this occurred in 1980 when the city of Brandon joined with the adjacent rural municipalities of Elton and Cornwallis to adopt and administer a joint planning program (Chapter 8).

Throughout the past seventy years, planning has become an increasingly complex process. The complexity of the various aspects of urban development has resulted in the emergence of a more sophisticated approach to planning — a system of "comprehensive planning" which recognizes and responds to the complex nature of urban functions and growth processes. The comprehensive planning process has evolved as a very analytical and broadly based process, which takes social and economic factors into consideration along with the traditional physical factors. This process has become more generalized in terms of long-range physical plans (master plans), and has placed more emphasis on policies and programs to manage the process of urban development. Planning has thus become an increasingly important function of local government. Planning policies, programs and decisions are, to a significant extent, political policies and decisions, and are therefore significantly influenced by the perceptions, priorities and personalities of elected officials.

## A Planning Perspective of Brandon's Development

A number of factors have contributed to the present pattern of urban development in Brandon. One of the most important was the Dominion government survey of the Prairies into the township system during the 1870s, which greatly influenced the pattern of subsequent development. The township system established sections and quarter sections of land, and a grid of road allowances ninety-nine feet in width. Many of Brandon's present arterial streets, such as First, Eighteenth and Thirty-Fourth Streets, Richmond Avenue, Victoria Avenue and Braecrest Drive, occupy these road allowances. A number of quarter sections of land were subsequently surveyed in the rectangular gridiron pattern, typified by lots with a depth of 120 feet and streets with a width of sixty-six feet. This gridiron system was easily and quickly surveyed, and it facilitated a speculative wave of subdivision not only within the city but also beyond its boundaries. These gridiron surveys were to determine the pattern of urban development in Brandon for years to come, particularly in the area between First Street and Twenty-Sixth Street north of Richmond Avenue (Figure 4:4).

Another important factor was the selection of the site for the station grounds by the CPR in 1881. This determined the location of the city's nucleus, and provided Brandon with its link to the outside world. It was only natural for early commercial and industrial development to be clustered adjacent to the station grounds, as this means of transportation provided an essential service for industry and commerce. In the absence of automobiles, early residential development tended to be clustered within walking distance of the centre of the city (Chapter 4). This resulted in a fairly dense pattern of development, typified by 1.5- to 2-storey homes on small lots measuring 25 feet by 120 feet. The development of the Canadian Northern Railway in 1906 further complicated the pattern of urban development. The railway diagonally traversed a number of pre-existing blocks and streets, thereby creating a number of small, oddly shaped fragments of private land, many of which were too small to be of much use, and a number of street crossings which presented traffic conflicts. The subsequent industrial development which occurred adjacent to this railway resulted in a mixture of land use in the midst of residential areas which persists to this day (Chapter 4).

Brandon experienced a period of vigorous development between 1880 and the onset of World War I in 1914. From its initial nucleus, the city expanded in a rather concentric manner in most directions (Figure 4:2). Development to the north was not as vigorous as in other areas of the city, due to the hazard of flooding along the Assiniboine River and the problem of river and railway crossings (Chapter 2). Several developments at the outskirts of the city around the turn of the century were to have a long-term influence on the city's future development. The establishment of Brandon College adjacent to Eighteenth Street, a hospital and a jail east of First Street, and the Exhibition Grounds in the south end of the city provided the seeds for the development of major institutional complexes during future years. In addition, these areas established growth points which attracted new residential development. These growth trends were reinforced by the creation of a streetcar system in 1913 which linked all of these far-flung developments, and encouraged infill development along its routes. Over time, the automobile replaced the streetcar as the prevalent means of mechanized transportation, and residential developments became more dispersed from the original pattern.

During the years of the Great Depression and World War II, the growth and development of the city came to a virtual standstill. In constrast, the postwar period was an era of unprecedented prosperity, which was reflected by the extensive growth and development of the city. The widespread use of the automobile, combined with the popular preference for bungalow housing, resulted in new subdivisions which were developed at a much lower density than the older, central areas of the city (Figure 4:4).

This decentralization trend was typical not only of residential development; it characterized industrial and commercial development as well. In establishing the Green Acres Industrial Park in 1962 in the southeastern portion of the city (Figure 4:5), council made a deliberate attempt not only to entice industrial development with the lure of inexpensive land, but also to segregate future industrial development from residential areas. A distinct trend of commercial decentralization has also occurred, with new commercial development being located along major arterial streets, particularly First Street, Eighteenth Street and Victoria Avenue (Chapter 7). This trend can be attributed to a number of factors, including the increased popularity of automobile transportation, the perceived problems of parking, traffic congestion and land assembly in the downtown area, modern merchandising preferences for large floor areas and on-site parking, and the comparatively lower cost of land located along the arterial streets.

Urban growth was not always confined to the area within the city limits. New development was enticed into the surrounding countryside in the rural municipality of Cornwallis (Chapter 8). During the mid-1930s, an oil refinery was established south of Richmond Avenue and east of Eighteenth Street (Plate 6:1). Residential development also overflowed the city boundary into the area south of Richmond Avenue. As well, various commercial and industrial developments arose in the area between Braecrest Drive and the Trans-Canada Highway (Chapter 8). The Simplot Chemical Company established a fertilizer factory east of Seventeenth Street East in the mid-1960s (Chapter 6). Such developments responded to the lure of cheaper land and lower municipal taxes in Cornwallis. While such developments were apparently welcomed by the rural municipality, they became a growing source of concern for the city because they benefitted from a number of facilities and services provided by the city, but did not contribute to the city's tax base. Unplanned industrial development beyond the city's boundaries created some problems for its residents. Also, development in these areas did not conform to urban development standards. This was particularly apparent in the residential area south of Richmond Avenue, which still exhibits a number of gravel surfaced streets and open drainage ditches. Until recently, parts of this area did not even have the benefit of city sewer and water services.

As a result of these problems, the city annexed approximately eighteen sections of land from Cornwallis in 1971 and 1972, in order to gain control over the development of land and the tax benefits generated by such development (Figure 8:2). The rural municipality was greatly opposed to this annexation, as it represented a substantial loss of land and tax revenue. This resulted in a period of acrimony between the councils of Brandon and Cornwallis.

Over the years, the development of the city has been influenced by a host of other factors, one of the most important being improvements in the highway transportation system. The vast improvements to the provincial highway system in the postwar era have greatly enhanced Brandon's role as a trading centre, as it became readily accessible to a large rural hinterland. Highway improvements within the city have also greatly influenced the pattern of urban development. The creation of the Trans-Canada Highway north of the city resulted in significant commercial developments along the highway and along First and Eighteenth Streets (Chapter 8). Improvements to Eighteenth Street, including the construction of the railway overpass, led to its establishment as the most prominent north-south arterial street in the city (Plate 7:4). With the increasing popularity of the automobile, Eighteenth Street and Victoria Avenue West emerged as important commercial strips.

## The Evolution of Planning in Brandon

Throughout Brandon's history, there have been a number of subtle planning processes. As previously indicated, many of these earlier processes involved locational decisions made by surveyors, engineers, architects, builders and property owners. Collectively, these decisions resulted in a pattern of industrial development along the railways and a cluster of commercial development in the downtown area. Industrial developers tended to locate close to the railways to satisfy their transportation needs, and commercial developers clustered in the downtown area in order to reap the financial benefits of a high-profile location within an established business area. Over the years, city council also made a number of planning decisions which to some extent influenced the growth of the city. Many of these decisions related to the location of new institutional buildings such as schools, and the provision of waterworks within various areas. Obviously such decisions had some influence on the direction of future residential development. However, there was no conscious effort to organize the overall pattern of the city's development until the postwar era.

### *Early Zoning in Brandon*

In 1948, city council adopted Bylaw No. 2954 which was a planning scheme containing the first zoning regulations for the city. This was a major step, as it signified the perceived public need for planning in order to promote the orderly growth and development of the city. The major purposes of the 1948 planning scheme were essentially twofold: to divide the community into zones for various types of development, and to establish development standards governing the density, size and position of buildings. The six zoning categories which were established consisted of the one- and two-family dwelling zone, the multiple-family dwelling zone, the local business zone, the general business zone, the industrial zone and the agricultural zone. To a large extent the arrangement of the zones was determined by the pattern of existing development at that time. The general business zone encompassed the downtown area and the industrial zone was located along the two railway systems. The one- and two-family dwelling zone encompassed much of the existing residential area. Areas for multiple-family residential development were scattered throughout the existing residential area, with a significant concentration around Brandon College (now Brandon University). The undeveloped portion of the city was zoned as an agricultural zone. The provisions of this bylaw also specified the types of development that would be permitted in each zone.

In addition to establishing separate zones, this planning scheme established setback requirements for all buildings, and minimum size requirements for residential lots. The building setback requirements varied from zero for commercial buildings in the downtown area, to a requirement for front setbacks of thirty feet and side setbacks of five feet for houses. In addition, this planning scheme prescribed a minimum lot size requirement of 5,500 square feet for each residential lot. This resulted in lot sizes with dimensions in the order of 50 feet in width and 110 feet in depth. This was a significant departure from the developments in the older areas of the city where typical lot sizes were 25 feet by 120 feet. This not only corresponded to the postwar trend towards low-density development and the construction of bungalow housing, but also signified council's intention to encourage new residential development in this fashion through the use of zoning requirements.

Several other aspects of this planning scheme are particularly significant. One of the general provisions prohibited the construction of buildings on land having an elevation of less than 1,180 feet above sea level, which in effect prohibited development on the floodplain of the Assiniboine River. In addition, the planning scheme required the developers of multiple-family dwellings to provide two hundred square feet of on-site parking area per family (which constitutes approximately one parking space per family). Similarly, the developers of certain types of business in the downtown area were required to provide off-street loading and parking facilities — an apparent attempt to alleviate vehicular congestion on downtown streets.

The 1948 planning scheme was an important beginning, as it established council's authority and commitment to regulate the process of urban development. One of its major benefits was that it provided for the development of attractive residential areas, and protected these areas from the intrusion of industrial development.

However, this planning scheme contained several weaknesses, one of its major shortcomings being that it was not sufficiently detailed to regulate the complex components of urban development. The provision of only one industrial zoning category did not make any major distinction between industrial developments which might be reasonably compatible with other types of development and industrial developments which might present problems. With only two residential zoning categories, it did not establish an appropriate range of provisions for the various forms of residential development. However, these shortcomings were eventually overcome.

The 1948 planning scheme was subsequently replaced by Bylaw No. 3383, The City of Brandon Planning Scheme 1958, which built upon the foundation established in 1948, and provided a more comprehensive system of zoning regulations which overcame the shortcomings of the previous bylaw. One of the most striking aspects of the new planning scheme was the increased diversity of zoning categories, which was more appropriate for a rapidly growing city. In contrast to the 1948 scheme, which provided only six zoning categories, the 1958 scheme provided sixteen zoning categories. This included six residential zones which provided for a range of residential development, from unserviced homes on large lots to multi-storey apartment buildings. It also provided for three industrial zones with provisions for light industrial zones to accommodate those types of industrial development which would be reasonably compatible with nearby development, and heavy industrial zones which were intended to accommodate those industrial developments with objectionable characteristics. The 1958 scheme also provided four commercial zones. The commercial zoning in the downtown area provided for the greatest diversity of commercial development, and the deliberate absence of setback requirements provided for the continuation of high density development in the downtown area. As well, the zoning provided for the continuance of industrial uses on the upper floors of buildings in the downtown area. The planning scheme also recognized emerging commercial trends by providing a "C1" zone for neighbourhood commercial development and a "C3" zone for highway commercial development. In theory, the neighbourhood commercial zones were intended to provide for convenience stores at the neighbourhood scale, and highway commercial zones were intended to provide goods and services to the travelling public, typified by establishments such as service stations and motels. This planning scheme also established zones for floodplains, parks and recreational areas, and institutional areas, as well as providing regulations for the development of these areas. It also responded to the increasing spatial requirements of the automobile, as it contained more stringent requirements for on-site parking, particularly in residential and commercial zones. While multiple-family buildings were still required to provide one parking space per dwelling unit, single-family dwellings and duplexes were now compelled to provide on-site parking to the same standard. All commercial developments were required to provide on-site parking, with the number of spaces being determined by a formula based on the floor area of the building. These requirements were not retroactive for existing buildings, but applied to new construction.

The 1958 planning scheme governed the growth and development of Brandon for twenty-five years, a period of substantial urban development. It provided a high degree of precision in its organization and regulation of development within the city. While it built upon the strengths of the 1948 planning scheme, it also contained some weaknesses. Its major shortcoming was the treatment of the commercial zoning categories. One of the major criticisms was that the downtown commercial zoning continued to encourage various types of industrial development in the downtown area, thereby perpetuating a number of visual and functional problems, rather than encouraging industrial development to relocate in more appropriate areas. More significantly, however, the neighbourhood and highway commercial zones permitted a number of traditional downtown businesses. This resulted in a proliferation of businesses along arterial streets, and indirectly contributed to a lack of commercial redevelopment in the downtown area.

*Comprehensive Planning in Brandon*

During the 1970s, Brandon embarked on a comprehensive planning program. An extensive study and analysis of the structure and dynamics of the city was undertaken in the mid-1970s and subsequently published as *Background Studies — Brandon Development Plan* (Province of Manitoba 1976). This report contained an extensive overview and analysis of many aspects of Brandon, including the physiography of the city, social and economic trends, housing and commercial trends, as well as the condition of the urban infrastructure (including sewers, waterworks and roads). *Background Studies* provided city council and the general public with an extensive body of knowledge on which to build the city's long-range development strategy.

The *City of Brandon Development Plan* (Province of Manitoba 1977) was subsequently prepared and adopted by city council as Bylaw No. 4738/15/77 in 1979. It specifies the city's long-range development objectives and a series of strategies which are to be used as a means of achieving these objectives. The *Development Plan* is intended to form the organizational basis for a number of other programs, such as land asssembly, capital works programs, and zoning regulations. The strategies contained in it are intended to respond to a broad range of development issues, with some of the more important aspects being discussed in the following paragraphs.

The *Development Plan* contains a number of planning strategies for the development of new residential areas in the city. One of the most significant of these stipulates that a mixture of housing is to be provided in new residential subdivisions. This provides for clusters of duplexes, townhouses and low-rise apartment buildings within new areas, with the objective of establishing a general mixture of socio-economic groups throughout the city. In addition, it identifies a number of planning principles for the development of new residential subdivisions, such as the provision of schools, parks and suitable street systems.

The *Development Plan* also supports the land assembly programs of the city, as well as the programs of the Canada Mortgage and Housing Corporation (CMHC) and the Manitoba Housing and Renewal Corporation (MHRC). The purpose of these programs is to ensure that there will continue to be an adequate supply of land for residential development. Another planning strategy is to provide for high-density residential development (high-rise apartments) in the vicinity of the downtown area, the objective being to provide for a large population in close proximity to the downtown area in order to bolster business and employment.

In terms of future commercial development within the city, the major objective of the *Development Plan* is to revitalize the downtown area. It contains a number of interrelated planning strategies for the redevelopment of major portions of the downtown area, and includes strategies for the preservation of heritage buildings, improved landscaping, and increased parking facilities on the periphery of the downtown area. It also prescribes a fifteen- to twenty-year moratorium on the construction of new suburban regional shopping malls, in order to channel commercial development into the downtown area and to provide sufficient time for revitalization efforts to succeed.

In terms of future industrial development, the main strategy of the *Development Plan* is to perpetuate the trend of the past twenty years. The most easterly portion of the city is designated for future heavy industrial developments which may generate significant amounts of noise, smoke, dust, or other emissions. The land immediately east of, and adjacent to, existing residential areas is designated as restricted industrial, and is intended to accommodate light industrial uses which would be compatible with adjacent residential development. The land between the heavy industrial and the restricted industrial areas is designated as general industrial, which provides for a range of intermediate types of industrial development. The *Development Plan* also provides for impact studies to be undertaken in the case of any proposed industrial development which might create an unacceptable level of emissions, and it allows performance standards to be established by council for those types of industrial development which may otherwise create unacceptable problems.

The *Development Plan* also contains a number of strategies for other types of development within the city. For example, it indicates that the floodplain of the Assiniboine River is to be developed as a recreational-open space corridor. In terms of transportation planning, it establishes a hierarchical street system, and designates a pattern of present and future collector and arterial street systems for the entire city.

The *Development Plan* was followed by the preparation and adoption of a more up-to-date set of zoning regulations in 1983, The City of Brandon Zoning Bylaw No. 5128/14/82. This new bylaw retains the traditional zoning control over the location of new development as well as various development regulations (such as height and setback requirements), and is coordinated with the long-range planning strategies of the *Development Plan.* The new zoning represents a refinement of previous zoning regulations, and has removed some of the ambiguities and filled in some of the gaps which were apparent in earlier bylaws. It provides much more precision in the listing of uses which are permitted in each zone, and the development requirements for new buildings. In addition, it contains a number of other significant improvements. First, many of the undeveloped agricultural areas within the city are zoned as residential reserve zones or industrial reserve zones thereby signifying long-range development intentions. Second, a large area of land surrounding the downtown area is zoned for high-density residential development in order to provide for more apartment construction in this area. Third, the new bylaw permits a slightly higher density of residential development than previous bylaws by reducing the minimum requirement for lot width from fifty feet to forty feet, and the minimum area requirement from fifty-five hundred square feet to four thousand square feet (for single-family homes). Fourth, the new bylaw has increased the requirements for on-site parking, particularly for apartment buildings and commercial developments, in response to increases in car ownership and problems with on-street parking.

*District Planning*

As previously indicated, there has been a general trend towards regional or district planning throughout Canada and the United States during the past twenty-five years. This trend grew from the realization that a number of development issues and concerns were shared by urban and rural municipalities. Some of these concerns stemmed from the expansion of urban development beyond urban boundaries, which led to issues such as the preservation of productive farmland, and the intrusion of development which might interfere with traditional farming interests. Other concerns related to the provision of various facilities which serve both rural and urban residents alike, such as regional transportation and drainage systems, recreational facilities and waste disposal sites.

In 1980, the city of Brandon and the rural municipalities of Cornwallis and Elton agreed to form the BAPD (Figure 8:1), which heralded a new era of cooperation between these municipal corporations. A district planning board was established, consisting of three representatives from each of the municipal councils. Since that time, the primary purpose of the district planning board has been the adoption and administration of a coordinated planning program for the planning district. One of the first and most important tasks of the planning board was the preparation of a development plan for the entire planning district.

The *Brandon and Area Planning District Development Plan* (Province of Manitoba 1984) was adopted by the board of the BAPD as Bylaw 9/84. This development plan has two major components. One component consists of the *City of Brandon Development Plan*, which has been incorporated into the *District Development Plan*. The second component consists of a coordinated set of land use and development strategies for the rural area. One of the most significant strategies provides for the preservation of high-quality farmland, by limiting the type of development that could take place on such land. Another important strategy provides for the safeguarding of farming operations from the intrusion of development, such as a cluster of residential "acreage" lots, which might compromise the nearby farming operations. Other strategies deal with the preservation of other rural resource areas, such as high-quality wildlife habitat and gravel deposits.

The district planning board also provides a continuous review of various development issues and projects. As an example, the board played a major role in reviewing and making recommendations on the proposed highway changes in this area, particularly the proposed eastern access highway linking the city's eastern industrial area to the Trans-Canada Highway. The district planning board also administers a number of planning- and development-related functions within the planning district, the most prominent being the building permit and inspection functions.

## Prospect

As in the past, the future of the city will depend to a large extent on the fortunes of the agricultural economy. Other determinants of urban growth would be further development of the city's industrial base, further expansion of the city's service functions, such as education and health care, as well as the decentralization of government functions from Winnipeg.

Although Brandon has existed for over a century, city council has only attempted to organize the total pattern of urban development through urban planning over the past forty years. Council's early efforts to organize the urban growth process were primarily confined to zoning. Over the years, city council has played a more influential role through the development and administration of more appropriate zoning regulations, and in the 1970s through the development of a more comprehensive, long-range planning process. During this period, other city initiatives, such as capital works programs, the land assembly programs for the industrial park, the downtown shopping mall and the Kirkcaldy Heights residential area, have not only complemented the zoning, but have been major indications of council's intention to influence the development of the city. Although the processes of urban change occur relatively slowly, the various instruments of urban planning have influenced the growth and development of the city in a very profound and positive manner over the past forty years.

Continued commitment to the urban planning process will assure continued successful management of the processes of urban development. The long-range policies and strategies established in the *Development Plan* will assure the development of the city in an optimal manner. The prospects for industrial development are good, with an abundance of designated land in the eastern portion of the city, with the prospects of improved accessibility through the eastern access connection to the Trans-Canada Highway, and the proposed linkage to the provincial five-hundred-kilovolt power network. Although continued residential expansion is anticipated in westerly and southerly directions, current initiatives of city council and the Brandon School Division would suggest that the area north of the Assiniboine River is expected to experience significant residential growth during the next few years. As well, continued programs such as the Westbran Work Activity program will result in further improvements along the river, making it a more important recreational corridor through the city (Chapter 2).

While the organization of future industrial, residential and recreational areas is reasonably predictable, the future of commercial development within the city is much less certain. One of the most significant urban trends in Brandon is the continued decline of the downtown area and the proliferation of suburban commercial development along arterial streets. Until recently, all types of commercial development were perceived to be beneficial to the city, irrespective of location. The development of the Brandon Shoppers' Mall in the southern portion of the city was greatly welcomed at the time of its construction, as it not only diversified the commercial sector, but also provided Brandon with another "big city" feature. Over the years, however, it has become apparent that the market cannot sustain both a strong downtown commercial area and expanding highway commercial developments. This points to a very significant and pervasive urban dilemma — to what extent will city council allow the trend of commercial decentralization to continue, and to what extent will it bolster and revitalize the traditional downtown area?

One school of thought calls for a system of commercial containment, by rigidly controlling both the availability of suburban commercial land and the types of suburban commercial development permitted, thereby indirectly focussing commercial redevelopment into the downtown area. Another school of thought indicates that the marketplace must be unrestrained, and that commercial development must be allowed to respond freely to contemporary shopping desires and merchandizing techniques. These are powerful beliefs which are held not only by national and international corporations, but by a significant proportion of the general public. Thus, one of the most profound issues confronting the city is the future of the commercial sector. The outcome will be determined by the interplay of very powerful and influential market forces, and by the extent to which city council will attempt to influence these forces through its urban planning program.

In our system of democratic government, the decisions and policies of civic governments are largely based on how they perceive the attitudes and desires of the general public. In the final analysis, the future of the city will be determined by the desires and actions of its citizens. Through a continuing process of public information and debate, a long-range comprehensive planning program can be sustained, to provide for an urban environment which will continue to fulfill the needs of a growing urban population.

**References**

Gerecke, K. 1977. The history of Canadian city planning. In J. Lorimer and E. Ross, eds., *The Second City Book: Studies of Urban and Suburban Canada.* Toronto: James Lorimer and Co.

Goodman, W.I. and E.C. Freund. 1968. *The Principles and Practise of Urban Planning.* Washington: The International City Managers' Association.

Mumford, L. 1961. *The City in History: Its Origins, Its Transformations and Its Prospects.* New York: Harcourt, Brace and World Inc.

Province of Manitoba. 1976. *Background Studies: Brandon Development Plan.* Brandon: Province of Manitoba, Municipal Planning Branch.

Province of Manitoba. 1977. *City of Brandon Development Plan.* Brandon: Province of Manitoba, Municipal Planning Branch.

Province of Manitoba. 1984. *Brandon and Area Planning District Development Plan.* Brandon: Province of Manitoba, Municipal Planning Branch.

Russwurm, L. 1977. *The Surroundings of Our Cities: Problems and Planning Implications of Urban Fringe Landscapes.* Ottawa: Community Planning Press.

# TEN
# Brandon's Role as a Service Centre in Southwestern Manitoba

*Christoph Stadel and John Everitt*

## Introduction

It has been argued that "centrality is a basic principle in human affairs" (Brush and Gauthier 1968:1), and it appears that the prairie region is an excellent laboratory for empirical testing of classical themes and models of centrality. Looking superficially at its setting, Brandon's location "seems to satisfy most of the theoretical assumptions which were used in the development of the Central Place Theory and centrality models" (Stadel 1972:77). The region of southwestern Manitoba appears to represent an almost ideal isotropic plain with a relatively homogeneous spatial arrangement of geographical factors: a uniform topography; a generally low population with mainly uniform distribution, except for major nucleations associated with service centres of different hierarchical levels; a mainly agricultural economy with little manufacturing; and a relatively dense network of railways and roads.

However, upon closer examination this apparent spatial uniformity is broken in a number of ways: (1) The topography, although generally flat, exhibits a number of variations and "breaks" which are potential barriers to the expansion of service areas of central places. These "breaks" are the deeply entrenched river valleys, lakes and marshes, and hills with almost no permanent populations (Riding Mountain and Duck Mountain to the north, Turtle Mountain to the south, the Bald Head Hills to the east, and Pembina Mountain to the southeast). (2) Although the prevailing rural settlement system in the form of dispersed farmsteads and widely scattered service centres exhibits a considerable degree of regularity, the arrangement of service centres and thus of overall population distribution exhibits a linearity — initially along the railway lines, but more recently along major (paved) roads and highways, especially near larger centres. (3) The population of southwestern Manitoba is a mosaic of distinct cultural groups, some of which have favoured spatial nucleation; for example, Native groups in reserves, Ukrainians, Scandinavians, and Hutterites. Most recently, rural populations have shown a massing trend in the municipalities adjacent to the larger urban centres, mainly a result of "exurbanization." (4) Although the economy of southwestern Manitoba is predominantly based on agriculture, the region is not completely devoid of manufacturing and other employment opportunities (Chapter 6). Furthermore, agriculture and other economic activities do not offer the same opportunities throughout the region, and therefore some economic disparity exists between municipalities. (5) The centrality pattern and the spatial expression of service areas in southwestern Manitoba is affected to a considerable extent by a variety of administrative and political boundaries, most noticeably the provincial and international borders. Within Manitoba, Brandon is peripherally located in the southwestern corner bordered to the south by the United States boundary (95 kilometres away) and to the west by the Saskatchewan boundary (115 kilometres away). (6) The central place system, the functional role of Brandon, and the extent of its service area are overshadowed by the dominance of the primate city, Winnipeg, located about 200 kilometres to the east. To the west Regina, being smaller in size than Winnipeg and at a greater distance from Brandon (370 kilometres), does not encroach upon Brandon's role as a central place. To the north, Brandon has no major urban competitor, although for lower order functions Dauphin, at a distance of 160 kilometres, has developed a considerable regional service role. To the south, Minot, North Dakota, is a similar size to Brandon but has a considerably larger range and number of goods and services and has on occasions, at times of favourable exchange rates and customs regulations, extended its retail service area into southwestern Manitoba.

Figure 10:1 summarizes, schematically, the location of Brandon within the hierarchical network of service centres in southern Manitoba and the major highways that stimulate regional interaction, as well as topographic and political barriers.

## The Hierarchy of Service Centres in Southern Manitoba

In *Towns and Villages in Canada*, Hodge and Qadeer (1983:75) address themselves to the regional relations of settlements such as Brandon and observe:

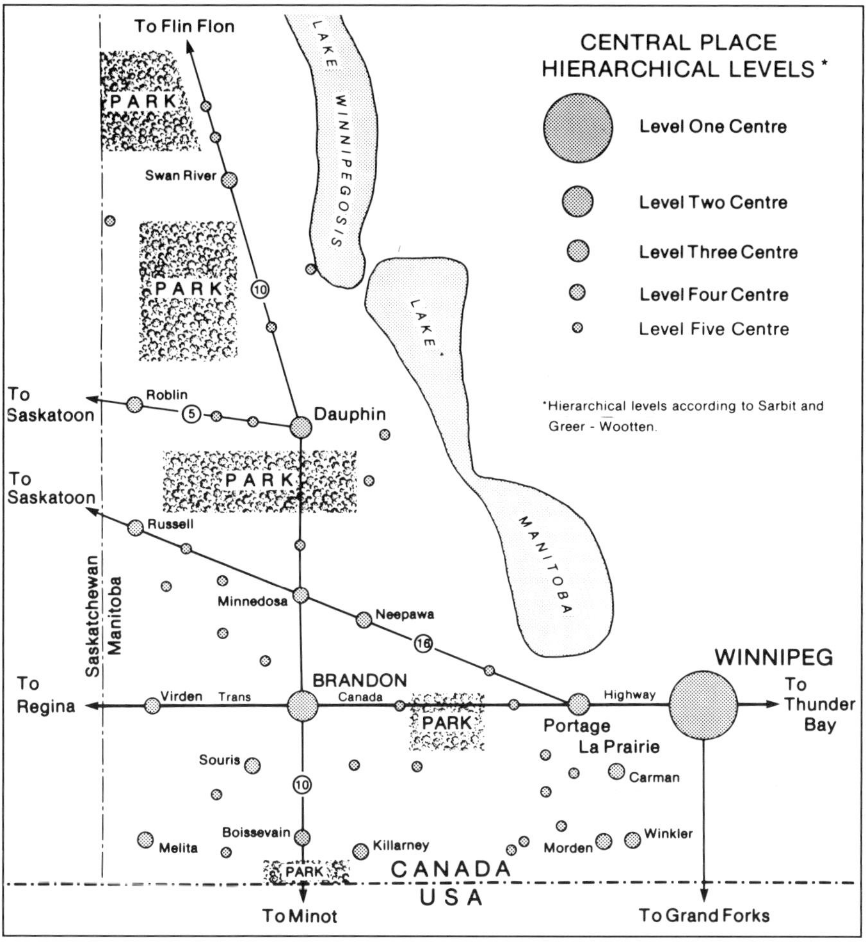


FIGURE 10:1
*Schematic representation of Brandon's location within the network of service centres in southwestern Manitoba*

Each small town exists within a larger national, provincial or metropolitan region. Since it seldom dominates such a region, the small town is subject to the forces and factors that govern the larger region's economy, social development, and growth prospects. But the small town is, at the same time, the center of its own subregion. . . . The small town, thus, has two major roles to play: it serves to organize the economic and social activities of its local region, and it participates with other towns and cities in

> the ongoing development of the large region. The popular wisdom is that the fortunes of the small town are entirely dependent on the fortunes of the larger region, of its economy and its cities.

Sarbit and Greer-Wootten (1980:100) found that "Southern Manitoba centres, by and large, conform to the spatial and non-spatial service structure proposed in central place theory" and that "it is not unreasonable to expect that the assumptions of the theory should be related to that system." It was also concluded that "the changes that have taken place in the system of centres in terms of growth and decline, conformed to generalizations made in the Manitoba government reports and also to results described in literature related to change" (p. 99). They also suggest that the primary indicators in the study of service centres, such as population numbers and changes, numbers and variety of goods and services offered, transportation networks, influence of the industrial sector, and impact of large cities, should be complemented by a set of contextual factors defined as "the environment of the system of central places" (p. 99).

Since the unpublished Ph.D. dissertation, "The Social and Economic Structure of the Trade Centres in the Canadian Prairie Provinces with Special Reference to its Changes 1910-1930" by Nathan L. Whetten (1932), various attempts have been made to classify, measure, and describe the different functional levels of the prairie community system. Zimmerman and Moneo distinguish four major categories of community: stop-off centres; home towns; farm cities; and prairie cities, and observe in their introduction that:

> These small communities tend to be organized into overlapping systems in which the commercial division of labor plays an important part. The larger towns and villages are subordinate, in a considerable extent, directly to the largest cities. . . . The very small villages are subordinate for most trading and services to the larger towns and small cities. . . . An open country family tends to have some trading, recreational or administrative relations with at least two or three nuclei (village, town, city), depending upon the services it wishes or needs. . . . Each family has near it a recognizable small village grocer (convenient for stop-off at); somewhat further away a

> high school centre (Home Town); further yet a weekly trade centre (Farm City); and finally the "City" (Prairie City). (Zimmerman and Moneo 1971:I)

According to Zimmerman and Moneo (1971:77), the typical "Total Farm City Community" on the Prairies would consist of one farm city (average population 3,300); three to four home towns (average combined population 2,150); eight to nine stop-off centres (average combined population 3,700); and 1,650 farms (average combined population 6,200). So, this "Total Farm City Community" would have a total average population of 15,350, 228 businesses, and would cover an area with a radius of twenty-five to thirty miles. In Manitoba in 1970, Winnipeg and Brandon were singled out as "Prairie Cities" and twenty-five communities were identified as "Farm Cities," with about half of them lying within southwestern Manitoba — the exact number depending on how the region is delimited. It was also found that the approximate distance between "Farm Cities" in Manitoba was thirty to forty miles, and that this spacing was considerably closer than in both Alberta (average distance, fifty to seventy miles) and Saskatchewan (average distance, fifty to sixty miles).

In an economic survey of the western Manitoba region, Brandon was classified as the exclusive "Central or Regional City" of the region supplying "a broad range of special services to most of the people in the region including higher education, specialized medical and professional services, a wider range of consumer items and farm supplies, recreational and cultural activities, radio and television services and a daily newspaper" (West-Man Regional Development Inc. 1968:29). There were eighteen "Shopping Centres" in the region providing "a wide range of goods and services through their 30 or more retail outlets. They [also] have modern high schools and hospitals, professional people, libraries, and . . . provide a variety of recreational facilities including golf courses" (p. 29). These places ranged in size from Erickson (1966:547 residents; 1981:528 residents) to Neepawa (1966:3,229 residents; 1981:3,216 residents). The third category identified was "Convenience Service Centres" with "smaller populations and with a narrower range of business and service functions than shopping centres" (p. 30). A total of nineteen such centres was counted in the West-Man area ranging in population from about two hundred (Kenton) to over five hundred (Elkhorn). Thirty-one communities fell into the next category, "Basic Service Centres," which provided "basic goods and services such as Grocery, Service Station, Drug Store, Hardware, Bank and Eating Places" (p. 30). These places ranged in size from about fifty (Glenora) to over three hundred (Oakburn). Finally, the lowest category was simply termed "Other Centres," whose "retail, service function is provided by two or three stores or less" (p. 31). The survey counted twenty-four communities in this category in the West-Man area. However, many of these places have ceased to have a service function, and all of them (except Forrest) have lost population since 1966.

Sarbit and Greer-Wootten (1980) distinguish six hierarchical levels of service centres in southern Manitoba without giving them specific names. Winnipeg, as the only Level One centre, and Brandon, as the only Level Two community, clearly dominate the urban hierarchy not only of southern Manitoba but also of areas to the north. At the third level, Portage la Prairie, and to a lesser extent Dauphin, "nest" within the wider service areas of Winnipeg and Brandon respectively. "Dauphin, located in a region with a relatively low density of sixth order places, tends to be more distant from centres supplying similar goods and services than Portage la Prairie, which is located between Brandon and Winnipeg, in a region of greater density" (Sarbit and Greer-Wootten 1980:43). Furthermore, Dauphin is "shielded" from Brandon's influence by the barrier of Riding Mountain National Park (Figure 10:1). Level Four centres in 1971 were concentrated more around Winnipeg (Selkirk, Beausejour, Gimli, Steinbach, Morris, Morden, Winkler, and Carman) than in southwestern Manitoba (Neepawa, Minnedosa, Virden, Souris, Killarney, Melita, Roblin and Swan River), and were also more closely spaced around Winnipeg. In southwestern Manitoba, a distinct lack of Level Four centres is noticeable in the area to the northwest of Brandon and southwest of Dauphin, possibly because this is an area of lesser economic strength, or it may be a result of the great drawing power of Brandon and Yorkton, Saskatchewan. Level Five places within the Brandon service area were Reston, Hartney, Deloraine, and Boissevain to the southwest and south; Elkhorn, Hamiota, Birtle, Shoal Lake, Rossburn, Russell, and Inglis to the west and northwest; Gilbert Plains, Grandview, St. Rose du Lac, McCreary, Minnedosa and Gladstone to the north and northeast; and Carberry, Glenboro, Holland, Treherne, Somerset, Cartwright, Crystal City and Pilot

Mound to the east and southeast. Level Six centres are not specifically mentioned in the Sarbit and Greer-Wootten study, but their number shown on Figure 3.4 of the study exceeds one hundred.

The Regional Analysis Program (RAP) Southern Manitoba examined and analysed the hierarchical pattern of service centres in southern Manitoba in two "Working Papers": *Analysis of Community Functions and Relationships* (Province of Manitoba 1974a) and *Analysis of Community Services and Facilities* (Province of Manitoba 1974b). In these papers, the settlements of southern Manitoba were classified into seven categories (Table 10:1). Winnipeg was excluded because of its excessive dominance in the settlement hierarchy of southern Manitoba.

TABLE 10:1
DISTRIBUTION OF POPULATION IN SOUTHERN MANITOBA, BY SETTLEMENT CLASSES: 1971

| Category of settlement | No. of centres | Percentage of total population of southern Manitoba |
|---|---|---|
| 1) City — Brandon | 1 | 8.2 |
| 2) Regional Centres | 5 | 10.5 |
| 3) Market Centres | 17 | 9.4 |
| 4) Local Centres | 54 | 9.7 |
| 5) Convenience Centres | 82 | 6.1 |
| 6) Stop-off Centres | 186 | 5.7 |
| 7) Single Enterprise Centres | 4 | 1.4 |
| Total of settlements of 50+ population | 349 | 51.0 |

*Source:* Regional Analysis Program Southern Manitoba (1974), Working Paper #2, *Analysis of Community Functions & Relationships*, (Province of Manitoba 1974b:3).

## Population and Functional Strength of Service Centres in Southwestern Manitoba

Table 10:2 portrays the population for 1971 and 1981 and the population change (1971-81) of the centres in southwestern Manitoba with a 1981 population of over four hundred. On the average, the relative strength of the service function of the centres is reflected by the size of the population change. Although there are a number of anomalies in the pattern of population change, the

TABLE 10:2
POPULATION AND POPULATION CHANGE OF CENTRES IN
SOUTHWESTERN MANITOBA: 1971-1981

| Population Centre | Population 1971 | Population 1981 | Percentage change 1971-1981 |
|---|---|---|---|
| Brandon | 31,150 | 36,242 | +14.0 |
| Portage la Prairie | 12,150 | 13,086 | + 2.6 |
| Dauphin | 8,891 | 8,971 | + 0.9 |
| Swan River | 3,522 | 3,782 | + 6.9 |
| Neepawa | 3,215 | 3,425 | + 6.1 |
| Virden | 2,823 | 2,940 | + 4.0 |
| Minnedosa | 2,621 | 2,637 | + 0.6 |
| Killarney | 2,074 | 2,342 | +11.4 |
| Roblin | 1,753 | 1,953 | +10.2 |
| Souris | 1,674 | 1,731 | + 3.3 |
| Russell | 1,526 | 1,660 | + 8.1 |
| Boissevain | 1,506 | 1,660 | + 9.3 |
| Carberry | 1,305 | 1,510 | +13.6 |
| Melita | 1,132 | 1,156 | + 2.1 |
| Deloraine | 961 | 1,135 | +15.4 |
| Rivers | 1,175 | 1,107 | - 5.8 |
| St. Rose du Lac | 818 | 1,090 | +25.0 |
| Grandview | 967 | 1,013 | + 4.5 |
| Gladstone | 933 | 964 | + 3.2 |
| Birtle | 882 | 887 | + 0.6 |
| Winnipegosis | 887 | 855 | - 3.6 |
| Pilot Mound | 763 | 838 | + 8.9 |
| Shoal Lake | 833 | 835 | + 0.2 |
| Gilbert Plains | 854 | 812 | - 4.9 |
| MacGregor | 744 | 795 | + 6.4 |
| Treherne | 628 | 743 | +15.5 |
| Glenboro | 698 | 741 | + 5.8 |
| Hamiota | 822 | 728 | -11.4 |
| Rossburn | 638 | 696 | + 8.3 |
| Minitonas | 610 | 628 | + 2.9 |
| Notre Dame de Lourdes | 613 | 627 | + 2.2 |
| McCreary | 545 | 618 | +11.8 |
| Somerset | 643 | 596 | - 7.3 |
| Erickson | 531 | 540 | + 1.7 |
| Elkhorn | 569 | 509 | -10.5 |
| Hartney | 579 | 490 | -15.4 |
| Wawanesa | 478 | 492 | + 2.8 |
| Crystal City | 555 | 489 | -11.9 |
| Ethelbert | 526 | 474 | - 9.9 |
| Bowsman | 443 | 454 | + 2.4 |
| Benito | 478 | 441 | - 7.7 |

*Source: Dominion Bureau of Statistics Census* 1971, 1981.

higher-order centres exhibit a larger increase, whereas the lower-order centres show a smaller increase or even a decrease. The table shows the continued primacy of Brandon within southwestern Manitoba, as well as the fact that it had the third highest population increase (+14 percent) between 1971 and 1981. The regional centres of Dauphin (1981: 8,971) and Swan River (1981: 3,782) exhibited increases of +0.9 percent and +6.9 percent respectively, whereas the nine market centres in southwestern Manitoba had an average increase of 6 percent, ranging from a high of +11.4 percent for Killarney to a low of +0.6 percent for Minnedosa. The population change of the twenty-eight local centres in southwestern Manitoba was an increase in some cases (St. Rose du Lac:+25.0 percent; Treherne:+15.5 percent; Carberry:+13.6 percent; McCreary:+11.8 percent) and a significant decrease in others (Hartney:-15.4 percent; Crystal City:-11.9 percent; Hamiota:-11.4 percent). From these population figures it becomes evident that the size of the settlement is not as important for the viability of a centre as one might assume. Also, there does not appear to be any clear indication of subregions with a distinct pattern of recent population change (Figure 10:2).

The Regional Analysis Program showed that the activities with the highest frequency of occurrence in southern Manitoba were churches (891), restaurants/taverns (857), general stores (629), trade contractors (595), and gasoline service stations (588); those with the lowest frequency of occurrence were universities and technical colleges (1). In a later study using 1981 census data and material contained in *Community Statistics* (Province of Manitoba 1985a), Caldwell (1985) found a slightly different hierarchy for settlements of over 500, the most frequently occurring services being restaurants/taverns (200), trade contractors (198), clothing/dry goods stores (168), food stores (156), gasoline service stations (145), and churches (140). In terms of the threshold population required the lowest values were noted for churches (21), restaurants/taverns (130), trade contractors (280), gasoline service stations (304), and elementary/secondary schools (336).

In the Regional Analysis Program, the relative strength of the centrality function of service centres was measured by population size, number and diversity of services, "distinguishing activities," the average "threshold population," and a "functional index." The relative strength was also determined by the availability of infrastructures which were grouped into the three categories "Basic

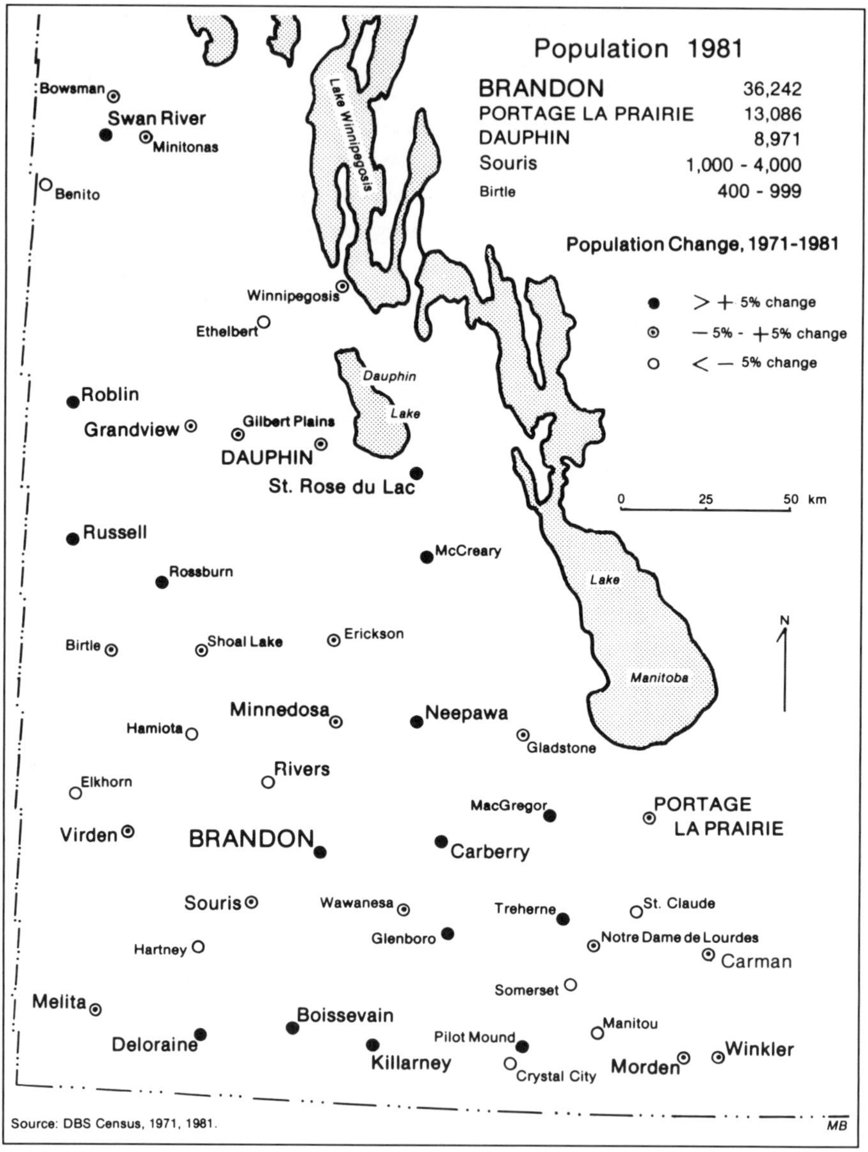


FIGURE 10:2
*Population change in the service centres of southwestern Manitoba: 1971-1981*

Infrastructure," "Social Infrastructure," and "Service Infrastructure" (Province of Manitoba 1974b:18-64). Each set of infrastructures was measured by an index which allowed for a comparison between centres.

Caldwell (1985) attempted to provide an update for 1981 of the available services and infrastructures of communities within Brandon's immediate trade area using the yearly *Report on Manitoba Communities* (Province of Manitoba 1971-1982), the *Manitoba Vacation Guide* (Province of Manitoba 1985b), the *Manitoba Telephone Directory* (Manitoba Telephone System 1985) and personal research data, but some of the data needed for a complete update of the Regional Analysis Program was unavailable. However, the information obtained confirms the continued dominance of Brandon, both in services and infrastructures in southwestern Manitoba. "Brandon appears to be the only place with truly 'urban' services or characteristics, while the remaining large places resemble large farm or rural settlements in comparison." However, "Winnipeg is so thoroughly and completely in control of the entire system, so much larger in every way than Brandon, that, in a total system perspective, Brandon is more like the remaining places than Winnipeg" (Sarbit and Greer-Wootten 1980:63-64).

## The Service Areas of Brandon

Brandon's trade area can be subdivided into a number of distinct service areas with a gradual weakening of Brandon's influence with increasing distance from the city. In *Background Studies: Brandon Development Plan* (Province of Manitoba 1976), a distinction is made between a primary trade area, a secondary trade area, and a tertiary trade area, based on a survey of owners/managers of industrial, commercial and service outlets in the city (Figure 10:3).

The primary trade area "is the region which is served on a regular and intensive basis by Brandon. Forty percent of the businesses surveyed (107 establishments) had trade areas which, when averaged, yielded this zone of influence. Six government offices and one educational office also served the area" (Province of Manitoba 1976:2-32). The primary trade area is also Brandon's major commutershed although commuting into Brandon for work is of relatively little importance (Brezden 1984). The area, which is almost circular in shape and has a radius of about fifty kilometres (forty to

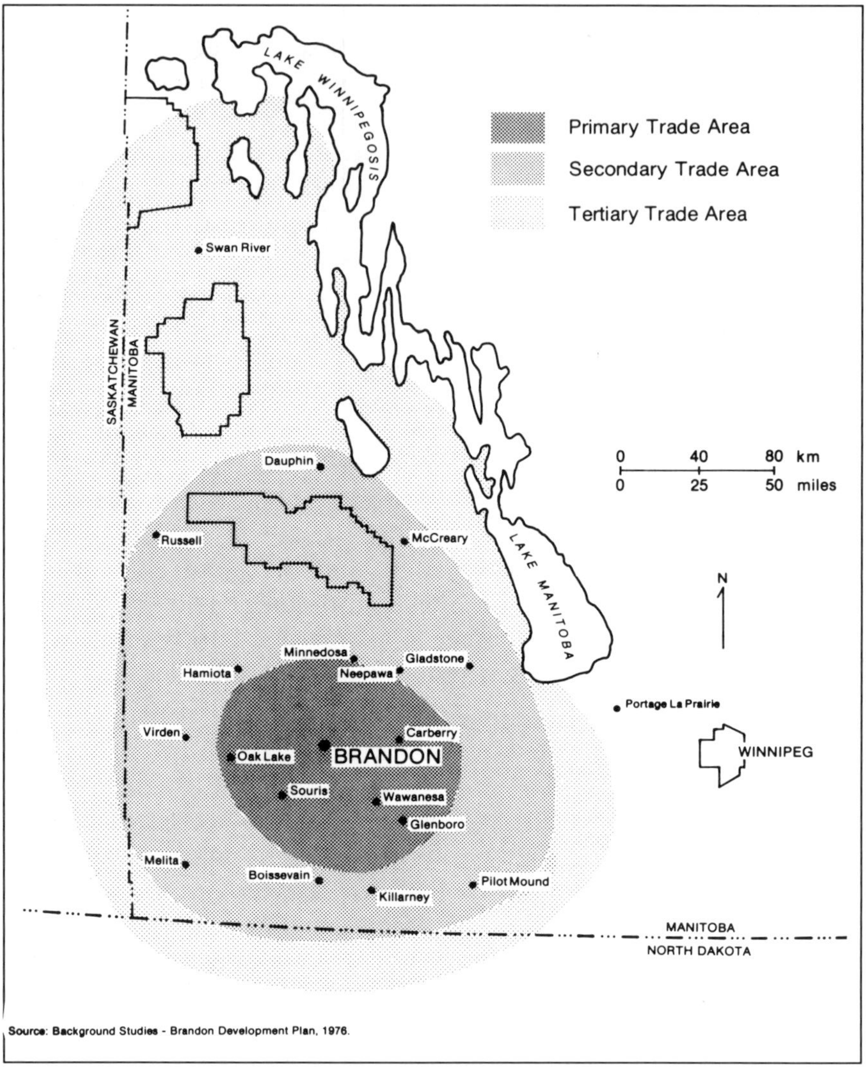


FIGURE 10:3
*Brandon's service areas*

fifty minutes travel time), includes, in addition to Brandon, nine incorporated settlements (1981:45,750 people), ten entire rural municipalities (1981:12,715 people) and eleven partial rural municipalities (1981:estimated 6,350 people) for a total of close to 65,000 people.

The secondary trade area, an area of less intensive and less frequent contact with Brandon, was considered by 34 percent of the businesses and nine government offices surveyed as their service area (Province of Manitoba 1976:2-33). It is elliptical in shape and finds its greatest extension to the north, extending to and including Dauphin (160 kilometres distant). To the west and south it is delimited by the provincial (115 kilometres distant) and international (95 kilometres distant) boundaries respectively, and extends eastward along the Trans-Canada Highway to a line between MacGregor and Portage la Prairie (about 100 kilometres), for a travel time distance varying from about seventy-five minutes to two hours. The secondary trade area includes thirty-eight incorporated settlements (1981:81,758 people), forty-four entire rural municipalities and the local government district of Park (1981:51,697 people), as well as fifteen partial rural municipalities (1981:estimated 10,900 people) for a total population of 144,355 people.

The tertiary trade area also shows its greatest extension to the north to Porcupine Mountain (about 420 kilometres distant) To the west it extends some 50 kilometres into Saskatchewan, to the south about 30 kilometres into North Dakota, and to the east it includes Portage la Prairie (126 kilometres). The travel time distance varies from ninety minutes to five hours. As this trade area shows a rather ill-defined extension into Saskatchewan and the United States, it is difficult to estimate its total population, but it must be close to 200,000 people.

Although the influence of a number of businesses and services (for example, the university) extends considerably beyond the limits of the tertiary trade area, especially to the north, no attempt was made in the *Background Studies* to portray and delimit a further extensive service area. However, it is noted that although "Brandon's trade area has developed most extensively in a northward direction . . . because of the lack of another major competitor in northwestern Manitoba . . ., the potential of northward expansion is also limited because of the sparse population in Northern Manitoba, the lack of commercial air transportation out of Brandon and the role of Winnipeg as a supply centre for Northern Manitoba" (Province of Manitoba 1976:2-33).

Since 1976, this three-tier delimitation of Brandon's trade area has been adopted in most reports and studies, including the recent *Brandon and Area Community Economic Development Study* (Local Employment Assistance & Development Program (LEAD Corporation) 1985). A number of unpublished student papers suggest that little change has occurred in recent years in the extent of Brandon's service areas, except for the tertiary trade area. Probably, the greatest change has occurred in the south where, because of monetary and other economic developments, the international border has become a sharp limit even for the tertiary trade area. In the other direction it appears reasonable to extend the tertiary trade area for a number of higher-order functions, notably the university, further to the north to include both The Pas and Flin Flon. The extension of the tertiary trade area to the east beyond Portage la Prairie appears problematic because of the powerful central role of Winnipeg and even the strengthened influence of Portage la Prairie resulting from a new shopping mall. To the west, on the other hand, the extension of Brandon's tertiary service area into eastern Saskatchewan appears realistic, a change that probably results from the construction of the new downtown mall and the expanded suburban Brandon Shoppers' Mall (Chapter 7).

In summary, Brandon's centrality function and the extension of its service area into a wide surrounding region is stimulated by the overall low density of a rather affluent population, by the absence of nearby major competing service centres and manufacturing plants, by a relatively level and open topography, and by the mobility of the population resulting from large-scale motorization coupled with a dense road network both of which tend to shrink real distances in the perception of the population.

On the other hand, a number of factors constrain Brandon's central role and are responsible for the delimitation of the service areas (Figure 10:1). To the north the "mountains," foremost among them Riding Mountain, hinder accessibility and have contributed to the strong function of Dauphin as a regional service centre. To the northeast, another effective barrier is provided by Lakes Manitoba and Winnipegosis, and to the east Spruce Woods Provincial Forest and Park restrict at least the intensive primary trade area.

In terms of competing centres, it has been shown that the overwhelming dominance of Winnipeg and the strong regional service function of Portage la Prairie have significantly restricted Brandon's reach in an easterly direction. Although Regina to the west and Minot, North Dakota, to the southwest are competing centres for Brandon, because of their greater distance they have not greatly influenced the "expected" extent of the city's service area. Yorkton, Saskatchewan, however, appears to erode Brandon's central function and service area northwest of the city close to the western part of Riding Mountain.

A third set of constraints are the major administrative and political boundaries which restrict Brandon's influence at least for some functions which are sensitive to these boundaries (e.g., Brandon University, Assiniboine Community College). This is, of course, particularly the case with the provincial boundary to the west and the international boundary to the south, but it also applies, to a lesser degree, to regional administrative boundaries for provincial services (e.g., planning districts, hospital districts, school districts).

Thus Brandon's service area appears to be delineated by different factors: the political boundaries to the west and south; physical features to the north and northeast; and competing centres to the east, southeast and northwest (Figure 10:1).

### *Brandon's Role as a Commercial and Industrial Service Centre*

Brandon's commercial function and its role as a regional shopping centre for southwestern Manitoba and the southeastern fringes of Saskatchewan is possibly the most important facet of the city's regional service function. Already in 1971, the *Data Analysis Program Southern Manitoba* (Carvalho-Page Group 1971:7) reported that an estimated 50 percent of Brandon's employment was service and trade related. According to the *Financial Post* (1984:80), Brandon's total retail sales volume reached $24 million in 1983 with per capita sales for that year amounting to $5,726, the highest value of all census divisions in Manitoba (provincial average, $3,846) — even higher than for Winnipeg ($4,426). Even at the national level, of the 23 centres in the 30-40,000 population bracket, Brandon had the fourth highest per capita retail sales volume; in the 30-100,000 category, Brandon was eleventh ranked among 48 urban places. In 1974, Brandon accounted for 39.3 percent of the total retail sales volume of southwestern Manitoba (Province of Manitoba 1976:2-35). During the late 1960s, Brandon's share of the provincial retail trade was consistently under 5 percent, but it had climbed to slightly over 5 percent by the late 1970s (Freshwater and Strain 1982:15) and reached 6.1 percent in 1983 (*Financial Post* 1984:473).

Local retail businesses are complemented by subsidiaries of major chain and department stores. Also, the regional impact of Brandon's retail function has been strengthened by the Brandon Shoppers' Mall and its expansion, the K-Mart Plaza, and the Brandon Gallery downtown mall. As in larger urban centres, these malls were designed to serve not only the city population but a larger surrounding trade area.

According to the former manager of the Brandon Shoppers' Mall, at least 50 percent of the customers and sales volume of the mall originate outside Brandon. However, this was not totally confirmed by a survey which in the summer of 1977 showed that in a sample of 3,627 draws at the Brandon Shoppers' Mall 78.5 percent of the entry forms were from Brandon and only 21.5 percent from outside the city (Sturko 1978). The addresses listed on the entry forms allowed an approximate delineation of the service area of the Brandon Shoppers' Mall. Of the 21.5 percent of the shoppers from outside Brandon 19.2 percent were from southwestern Manitoba and southeastern Saskatchewan; the remaining 2.3 percent originated from other parts of North America. Based on these figures, Sturko constructed a four-level trade area for the Brandon Shoppers' Mall. The inner ring, the primary service area, coincides approximately with the extent of the exurban area of the city, and is an area of regular and almost exclusive shopping in Brandon. It is at a travel distance of about fifteen to twenty minutes, and accounts for 25 percent of the non-Brandon shoppers. It includes, to the west, the communities of Kemnay and Alexander; to the north and northwest, Forrest, Rivers, and Rapid City; to the east and northeast, Justice, Douglas, Chater, and Shilo; and to the south the exurban dispersed settlements in the Brandon Hills.

The secondary service area includes the zone of frequent shopping in Brandon (accounting for 50 percent of the total of regional shoppers) at a maximum travel distance of about thirty minutes. It extends in an almost circular form around Brandon and includes the communities of Oak Lake and the Sioux Valley Indian Reserve to the west, Minnedosa to the north, Carberry to the east, Wawanesa to the southeast, and Souris to the southwest.

The tertiary service area (accounting for 75 percent of the shoppers in the trade area) is the zone of occasional shopping. It also has a circular form and can be delimited approximately by the sixty minute isochrone of travel time. It includes Virden to the west, Shoal Lake to the northwest, the southern edge of Riding Mountain National Park to the north, Neepawa to the northeast, Sidney to the east, Glenboro to the southeast, Boissevain to the south, and Melita to the southwest.

The quaternary service area, accounting for the rest of the shoppers in the trade area, is a rather vague zone of infrequent shopping in Brandon. Its outer boundary is diffuse and difficult to establish, but could be delimited approximately by the 90-120 minute isochrone of travel time which extends the outer service area of the Brandon Shoppers' Mall west to Moosomin in Saskatchewan, to Rossburn and Russell in the northwest, to Dauphin in the north, to Portage la Prairie in the east, to Pilot Mound and Crystal City in the southeast, and to the extreme northern part of North Dakota in the south.

The largest number of customers at the Brandon Shoppers' Mall came from Rivers, Shilo, Minnedosa, Virden, Souris, Wawanesa and Carberry, reflecting their proximity to Brandon, the size of the communities, and the quality of the connecting highways (the Trans-Canada Highway and Highway 10). The service area of the Brandon malls may be restricted, however, by the recent emergence of shopping malls in smaller centres located within the wider service area of Brandon; for example, the malls in Portage la Prairie, Dauphin, Killarney, and Virden.

Kinnear (1984) examined the consumer image and area of influence of both the Brandon Shoppers' Mall and the downtown Brandon Gallery. Based on a sample of 60 respondents from outside the city, Kinnear found that 82.5 percent of the shoppers within a 50 kilometre radius of Brandon and 30 percent of the customers outside the 50 kilometre radius make one or two trips per week to the Brandon Shoppers' Mall. In contrast, only 50 percent of the customers inside the 50 kilometre radius of Brandon and 28.6 percent of the customers outside the 50 kilometre radius shop once or twice a week at the Brandon Gallery. In comparing the two malls in terms of the strength of their regional pull, it appears that the gallery draws shoppers from a smaller area; this is particularly true for the southern and eastern parts of Brandon's service area.

In a more general delimitation of Brandon's service area for retailing, Stadel (1972:87) distinguished an intensive service area and a fringe service area. The intensive service area was almost circular with a radius of 30-40 kilometres, corresponding to a travel time of 30-40 minutes, and included a population of about 50,000. The fringe service area exhibited a north-south elongation extending northwards as far as Swan River and southwards into the northern parts of North Dakota. In a west-east direction, this fringe area was greatly restricted by the competing influence of Winnipeg to the east, and by the provincial boundary and competing Saskatchewan centres (Broadview, Estevan, Regina) to the west. There is no evidence to suggest a major change from this delimitation in the last fifteen years.

Brandon's intensive linkage with its agricultural hinterland is manifested in numerous ways. As was demonstrated earlier, the relative health of the retail business in Brandon is closely related to the fate of agriculture in southwestern Manitoba. In addition, the three major types of industrial activity in the city — the chemical industry (Simplot, Ayerst Organics, Canadian Occidential Petroleum Ltd.), the metal and fibreglass fabrication industry (Canadian Steel Tank, Behlen, Cancade, Hamilton and Allen), and the food processing and related industries (Burns, McKenzie Seeds) — are directly related to Brandon's agricultural hinterland (Chapter 6). The same is true for a number of transportation, wholesale, and retail businesses, among them eight grain elevators, including Manitoba Pool's new high-throughput elevator, implement dealers and farm supply outlets. It also applies to many of Brandon's financial and professional services (real estate, legal, accounting, medical).

Brandon's agricultural orientation is also evidenced by the importance of the Brandon Livestock Market with a sales volume of $72 million in 1982 (Brandon Industrial Commission 1985:42). On a regular basis, the livestock market serves an area with a radius of about eighty kilometres except to the east where the influence of the Winnipeg Livestock Terminal in St. Boniface extends as far west as Carberry (forty-five kilometres from Brandon). The less extensive service area of the Brandon Livestock Market extends into eastern Saskatchewan, but to the north Dauphin is a major competitor. Brandon's tradition as an agricultural service centre is highlighted each year by the Royal Manitoba Summer and Winter Fairs which attract a large number of visitors and a large amount of business from the surrounding rural area.

### *Brandon's Role as a Public Sector Service Centre*

The public sector function of Brandon could be considered as a spinoff from Brandon's role as the major regional service centre of western Manitoba. Its importance as a regional administrative centre is enhanced by the fact that the city is the second largest urban centre in Manitoba and is located at a considerable distance

(200 kilometres) from Winnipeg, where public sector functions are concentrated. It required a political decision to decentralize public service functions before there was an increase in the number of regional offices in Brandon. Decentralization was spurred by a provincial regional development strategy in the early 1970s documented in the *Guidelines for the Seventies* (Province of Manitoba 1973).

Estimates by the Brandon Industrial Commission in 1982 suggest that the public sector in Brandon employs some five thousand people. Twelve of the twenty largest employers in Brandon fall into the public service sector, including various federal agencies represented by a regional office in Brandon as well as numerous branches of provincial departments and provincially supported institutions. Although the tradition of hosting government institutions goes back to the establishment of the Brandon Agricultural Research Station in 1886, many provincial and federal agencies have been added only recently, most of them housed in a Federal Building on Princess Avenue and in a Provincial Building on Ninth Street.

Military establishments, although not based in Brandon, except for the Armouries, have for several decades strengthened the retail sector of the city and have provided employment for Brandon residents. Whereas the Rivers Air Force Base was closed in the late 1960s, the CFB Shilo received a boost in the 1970s when it became a military training centre for the German Army.

It is beyond the scope of this study to outline the administrative boundaries and service areas for all the different public sector functions. Suffice to say that the boundaries of some of the public sector service regions appear rather illogical and/or arbitrary (e.g., the Brandon Hospital District boundary and the Brandon School Division boundary). Clearly the boundaries of some government districts and the locations of some regional government offices are the result of political lobbying and government policies.

In an earlier study of Brandon's service areas, Stadel (1972 100-102) delimited a number of selected administrative districts for Brandon. While most administrative regions are clearly delimited by well-established boundaries (e.g., the Hospital District), other public sector service areas (e.g., the area of influence of Assiniboine Community College and Brandon University) exhibit diffuse boundaries and overlap the zones of other urban centres. The extent of the administrative districts varies from very small areas around

Brandon (e.g., Brandon School District, Manitoba Hydro District), to large service areas (e.g., Water Control District, Mines and Natural Resources District). Whereas many public sector service areas are clearly delimited to the west and south by the provincial and international boundaries, delimitation of these service areas to the east and especially to the north is much more problematic. To the east, Brandon's public sector service area is delimited, and in some cases overlain, by Winnipeg's higher-order administrative service function. To the north, however, Brandon faces only the competition of lower-order centres, although Dauphin, The Pas and Flin Flon have developed their own centrality for a number of administrative functions.

Brandon University's zone of major influence extends westwards into the extreme southeastern part of Saskatchewan (the Broadview, Weyburn and Estevan areas), and southwards to the international border. To the east and southeast, the zone of attraction of Brandon University is greatly restricted by the encroachment of the larger Winnipeg universities — the University of Winnipeg and the University of Manitoba (which in addition offers graduate and professional programs that are not available in Brandon). Consequently, the zone of influence of Brandon University does not extend much beyond Killarney and Glenboro in the southeast, Carberry in the east, and Neepawa in the northeast. To the north, Brandon University is less directly affected by the competition of the Winnipeg universities although the major northern centre, Thompson, is clearly located within the Winnipeg service area. Brandon University's reach into northern Manitoba is hampered by the lack of direct air connections to northern centres, whereas such direct links exist between Winnipeg and northern communities. On the other hand, Brandon University has a province-wide service area for a number of programs, especially the School of Music graduate program and the specialized programs for Native students such as Brandon University Native Teachers' Education Program (BUNTEP) and the Program for the Education of Native Teachers (PENT), and as a result of its participation in the Inter-Universities North Program (IUN). Through the IUN and Native education programs, Brandon University courses have been taught in many centres and Indian reserves in northern Manitoba.

### *Brandon's Role as a Transportation and Communication Service Centre*

"Accessibility of a service centre determines to a large degree the extent of its service areas, especially in regions of relatively even population distribution. It is not only determined by physical conditions, such as relief, but increasingly by the hierarchical position of the centre and by the efficiency of the transportation system in the region. Brandon's role as a service centre within the urban system of southern Manitoba and eastern Saskatchewan depends greatly on the quality and efficiency of the regional transportation system" (Stadel 1972:90).

Although the railways played the most significant role in the foundation and early growth of Brandon and its function as a regional service centre (Chapter 4), the importance of this role has declined over the last few decades reflecting growing competition from road transportation. Although the city is serviced by the Via Rail passenger system on the CPR mainline and on the CNR mainline (through Brandon North railway station about fifteen kilometres from the city), most of the regional and national passenger transport is done by bus. The Brandon region is serviced by two bus companies operating on regular schedules: Greyhound Bus Lines and the Manitoba Motor Transit Company. The national and international Greyhound Bus system links Brandon with communities along the Trans-Canada Highway (eight buses daily eastbound and five buses daily westbound) and with major centres outside Manitoba. Greyhound Bus Lines also service the communities further south along Highway 2 between Winnipeg and Regina by establishing a connection by a feeder taxi. Communities to the north and northwest are served by Greyhound Bus Lines along the Yellowhead Route which passes through Russell, Minnedosa, Neepawa and Portage la Prairie. The subsidiary company, Webb Bus Lines, has a daily service between Brandon, Souris and Melita.

The Grey Goose Line, operated by the Manitoba Motor Transit Company, has a twice daily service to Dauphin, Swan River, The Pas, Flin Flon and Thompson; daily services to Winnipeg, Boissevain and Killarney, with a transfer route westwards to Melita and into eastern Saskatchewan; as well as weekday services in a northwesterly direction to Russell, and in a southeasterly direction to Wawanesa and Ninette. The network of regularly serviced bus lines from Brandon does not cross the international boundary via Highway 10 although in the 1940s the Manitoba Motor Transit Company made several attempts to establish a regular bus service to Minot, North Dakota. The service proved to be uneconomical and was discontinued.

Recently Brandon's airport became a regional "window" to major Canadian centres by offering a daily air service to Winnipeg, Thunder Bay and Toronto in the east, and to Calgary and Vancouver in the west.

Brandon fulfills an important role as a regional service centre for freight movements, express freight being moved by the two bus companies along their lines of service. In addition, twenty outside and five local trucking firms serve the city and southwestern Manitoba and have links to other points in North America (Brandon Industrial Commission 1985:23).

In terms of railway freight service, Brandon continues to play an important regional role. The city is serviced by both the CPR and the CNR companies which provide a combination of rail and truck services. Brandon continues to be a divisional headquarters for CPR and daily train service is provided to the CPR piggyback trailer terminal at Brandon. The city and surrounding region is also served by the CP Trucking Group through its sixteen-bay terminal at Twenty-Sixth Street. Finally, CNR provides a piggyback service as well as an import/export container service on request, as illustrated by the fact that an average of 210 railway cars are loaded and unloaded in the city by CNR every week (Brandon Industrial Commission 1985:25). This important regional transportation function reflects Brandon's favourable location on, or near, the two transcontinental railways, at the intersection of two long-distance, high-volume highways — the Trans-Canada Highway and Highway 10 — and within a major east-west air traffic corridor.

Brandon also fulfills a major regional media and communications service role. The city's two AM radio stations (CKX and CKLQ), an FM radio station (KX96), a TV station (CKX), a community TV channel (Westman Media Co-op), and a TV station originating in Winnipeg but with a repeater in Brandon (CKYB), serve not only the city but the entire region of southwestern Manitoba and southeastern Saskatchewan (Brandon Industrial Commission 1985:43). As is the case with other services, Brandon's audio-visual media service area finds a major competitor in the Winnipeg stations which broadcast within the entire southern Manitoba region. Also, unlike most other services restricted by the international border, the radio and TV stations of the United States reach well into southwestern Manitoba, especially through the cable TV system introduced to Brandon in 1978.

Brandon's daily newspaper, the *Brandon Sun*, with a circulation of about twenty thousand, mostly serves the city, but it is distributed throughout southwestern Manitoba. This is also true for the advertising weekly, The *Cross Country Shopper*, and the *Sunday Sunday* [*sic*] paper. In addition, since the *Globe and Mail* has its regional printing centre in Brandon, this national newspaper is distributed to the surrounding region. As was the case for the radio and TV stations, the Winnipeg daily newspaper, the *Winnipeg Free Press*, and the tabloid, the *Sun*, encroach into the service area of the *Brandon Sun*.

## Conclusion

This chapter demonstrates that Brandon's functions and viability are strongly linked to the surrounding area, and that the city's major role is that of a regional service centre for a wide trade area and a significant population:

> Brandon would appear to be developing, to a regional centre of a higher magnitude than expected, given its population size, industrial structure etc. This may be attributed . . . to the increased mobility of people in rural areas and smaller communities, as well as to increased effort by the government to make Brandon a regional centre. (Sarbit and Greer-Wootten 1980:147)

## References

Brandon Industrial Commission. 1985. *Brandon at the Centre*. Brandon: Brandon Industrial Commission.

Brezden, D. 1984. The City of Brandon as a Regional Centre: Brandon Commutershed. Student paper, Department of Geography, Brandon University.

Brush, J.E. and H.L. Gauthier. 1986. *Studies on the Philadelphia Metropolitan Fringe*. Chicago: University of Chicago, Department of Geography Research Paper 113.

Caldwell, G. 1985. The Centrality of Brandon in the Southwestern Manitoba Region. Student paper, Department of Geography, Brandon University.

Carvalho-Page Group. 1971. *Data Analysis Program Southern Manitoba.* Winnipeg: Province of Manitoba, Department of Industry and Commerce.

Dominion Bureau of Statistics, Census 1961, 1971, and 1981. Ottawa: Queen's Printer.

*Financial Post.* 1984. Canadian Markets 1984.

Freshwater, D. and G. Strain. 1982. *Drought Sensitivity Analysis Project: Brandon Study.* Winnipeg: University of Manitoba, Department of Agricultural Economics and Farm Management.

Hodge, G. and M.A. Qadeer. 1983. *Towns and Villages in Canada. The Importance of Being Unimportant.* Toronto: Butterworths.

Kinnear, M. 1984. Consumer Image of Brandon's Shopping Centres. Student paper, Department of Geography, Brandon University.

Local Employment Assistance & Development Program (LEAD Corporation). 1985. *Brandon and Area Community Economic Development Study.* Brandon: Brandon University, LEAD Corporation.

Manitoba Telephone System. 1985. *Manitoba Telephone Directory.* Winnipeg: Manitoba Telephone System.

Province of Manitoba. 1971-1982. *Report on Manitoba Communities.* Winnipeg: Province of Manitoba, Department of Economic Development and Tourism.

Province of Manitoba. 1973. *Guidelines for the Seventies.* Winnipeg: Province of Manitoba.

Province of Manitoba. 1974a. *Analysis of Community Functions & Relationships.* Winnipeg: Province of Manitoba, Department of Industry and Commerce, Regional Planning and Development Branch, Regional Analysis Program Southern Manitoba, Working Paper 2.

Province of Manitoba. 1974b. *Analysis of Community Services & Facilities.* Winnipeg: Province of Manitoba, Department of Industry and Commerce, Regional Planning and Development Branch, Regional Analysis Program Southern Manitoba, Working Paper 3.

Province of Manitoba. 1976. *Background Studies: Brandon Development Plan.* Brandon: Province of Manitoba, Municipal Planning Branch.

Province of Manitoba. 1985a. *Community Statistics.* Brandon: Province of Manitoba, Business Development Centre.
Province of Manitoba. 1985b. *Manitoba Vacation Guide.* Winnipeg: Province of Manitoba, Department of Tourism.
Sarbit, L.A. and B. Greer-Wootten. 1980. *Spatial Aspects of Structural Change in Central Place Systems: Southern Manitoba 1961-1971.* Toronto: York University, Geographical Monographs 4.
Stadel, C. 1972. Service areas of a non-primate city in the Canadian Prairies: the case of Brandon, Manitoba. In A.H. Paul and E.H. Dale, eds., *Southern Prairies Field Excursion, Background Papers.* Regina: University of Saskatchewan, Regina Campus.
Sturko, D. 1978. Service Areas of the Brandon Shoppers' Mall. Student paper, Department of Geography, Brandon University.
West-Man Regional Development Inc. 1968. *The Economic Base of West-Man.* Brandon: West-Man Regional Development Inc.
Whetten, N.L. 1932. The Social and Economic Structure of the Trade Centres in the Canadian Prairie Provinces with Special Reference to its Changes 1910-1930. Ph.D. thesis, Harvard University.
Zimmerman, C.C. and G. Moneo. 1971. *The Prairie Community System.* Ottawa: Agricultural Economics Research Council of Canada.

# ELEVEN
# Conclusion and Outlook

*John Welsted, John Everitt and Christoph Stadel*

In this volume we have attempted to paint a picture of Brandon during its first century, but what of the future? Predicting the future is dangerous but it is possible to project some trends, based upon the existing data. We will try to share some of our views on the future of Brandon to the year 2000 in order to serve as a final step in our analysis of the city.

It is obvious that major changes in landforms cannot be expected over the next twelve years. Hydrologically, the Assiniboine River is under reasonable control, which may have encouraged encroachment of both commercial and residential land use onto the floodplain. While the probability of flooding at Brandon has been reduced by the construction of Shellmouth Dam north of Russell and by the building of dikes within the city, the risk of flooding has not been completely eliminated. Buildings on floodplains are always at some risk even with the most elaborate defence precautions. However, the risk applies to only a small part of the city and even if flooding does occur it will have little effect on the city's overall urban fabric. The recent decision by the province to charge the city for water taken from the Assiniboine is a potential source of disharmony between the two government levels.

There is much discussion in the scientific literature of climate change with the suggestion that the climate of the Prairies is becoming hotter and drier — a trend which could have a profound effect on Brandon's future. Not only would it be more difficult to grow crops, the risk of soil erosion and soil degradation in southwestern Manitoba would be increased. While these problems are not as severe in southwestern Manitoba as in some areas further west, their adverse effects will become severe unless a well-organized program of soil conservation is practiced (Senate of Canada 1984). If soil conservation is not practiced, the future of agriculture in southwestern Manitoba, and thus Brandon's economic future, could be severely compromised.

Paradoxically, the socio-economic patterns in Brandon seem to be both settled and changeable at the same time. The "West End" has been clearly marked out as a territory for the upper status/upper income groups, and the "East End" continues its role as a region for working-class/blue-collar people. These generalizations are of course open to criticism if the situation is looked at in detail, but they do describe the major trends. In a similar vein, Assiniboine Gardens "north of the (CPR) tracks" is likely to remain noticeably Ukrainian and Polish in ethnic background in the near future despite recent changes caused by a minor influx of Native Indians and Anglo-Saxons.

However, north of Assiniboine Gardens the situation is more volatile. Some expensive houses have been constructed on the floodplain north of the river and enclaves of expensive, high-status housing are located on the North Hill, some of them only a stone's throw from mobile home developments. A new area on the North Hill west of Kirkaldy Heights (itself a very mixed social area) is being opened up by the city. It is a choice "view" site and has the added attraction of proximity to the Sportsplex and a new school. Given these facts, it is expected that the population of the North Hill will increase considerably in the near future. After more than a century the barrier of the Assiniboine is finally losing its significance.

In the "South End" the picture is unclear. In the late nineteenth century an area of scattered, large middle-class houses, this zone was adversely affected by the coming of the CNR, and the average value of land remained quite low. Now, however, the Brandon Shoppers' Mall and Crocus Plains Regional Secondary School have helped to generate growth, some of which consists of quite lavish homes, but some of which is inexpensive subsidized housing. Certainly there is not the uniformity of quality housing found in the "West End," but the "South End" has some advantages and parts of it may yet become mini "West Ends."

Manufacturing in Brandon will continue to depend on the city's agricultural hinterland. The lack of raw materials and the distance to potential markets preclude any major industrial development. There is always a chance that mobile high-tech industries will locate in the city taking advantage of its well-trained labour force, but such developments are impossible to foretell.

The planning situation for Brandon is probably the most volatile problem. The *City of Brandon Development Plan* (Province of

Manitoba 1977) has been changed repeatedly by later bylaws. Although recently a new effort has been made to save the downtown area, the city's policies and actions have favoured suburban development and have jeopardized the future of Brandon's core area.

The development and structure of the rural-urban fringe now seems to be better controlled, thanks to the *Brandon and Area Planning District Development Plan* (Province of Manitoba 1984). However, Brandon's fringe remains a highly dynamic and variable region with frequent and significant changes in land prices, land use, and ownership patterns as evidenced by the opening in 1987 of another motel on the Trans-Canada Highway next to a preexisting fast food outlet.

We expect a continuation of present trends with regard to Brandon's interaction with the southwest Manitoba region. The city will continue to serve as the region's principal cultural and commercial centre; the degree of primacy within the settlement network of southwest Manitoba may increase although there is no conclusive evidence of a major influx of rural people to the city.

In conclusion, although Brandon has in many ways reached a positive steady growth state in its first century, like all systems such a state may simply be a transitional position on the way to something quite different. There is, however, nothing to suggest that its primary role as a service centre for the agricultural region of southwestern Manitoba will change substantially in the foreseeable future.

## References

Province of Manitoba. 1977. *City of Brandon Development Plan.* Brandon: Province of Manitoba, Municipal Planning Branch.

Province of Manitoba. 1984. *Brandon and Area Planning District Development Plan.* Brandon: Province of Manitoba, Municipal Planning Branch.

Senate of Canada. 1984. *Soil at Risk.* Ottawa: Senate of Canada, Committee on Agriculture, Fisheries and Forestry.

## Notes on Contributors

**Paul Curtis** is the principal of J.R. Reid School in Brandon. He obtained a Bachelor of Arts from Brandon University in 1970 and a Master of Arts from the University of Alberta in 1972. He has been a sessional lecturer at Brandon University since 1972.

**John Everitt** is a professor of geography and chairman of the Department of Geography at Brandon University where he has taught since 1973.

**Alison Gill** is an assistant professor at Simon Fraser University. She taught geography at Brandon University from 1981 to 1984.

**Rod McGinn** is an assistant professor of geography at Brandon University where he has taught courses on weather and climate since 1979.

**Christoph Stadel** is a professor of geography at Brandon University where he has taught since 1968.

**Ron Tompkins** is a planner with the Municipal Planning Branch of the province of Manitoba based in Brandon.

**John Warkentin** is a professor of geography at York University. He has written numerous articles and books on the historical geography of the Prairies and was awarded the honorary degree, Doctor of Laws, by Brandon University in 1987.

**John Welsted** is a professor of geography at Brandon University where he has taught since 1965.